Water Resource Management and Flood Control Systems

WATER RESOURCE MANAGEMENT AND FLOOD CONTROL SYSTEMS

Carlos Andres Garcia Sanchez

ARCLER PRESS

www.arclerpress.com

Water Resource Management and Flood Control Systems

Carlos Andres Garcia Sanchez

Arcler Press
4164 Lakeshore Road
Burlington, ON L7L 1A4
Canada
www.arclerpress.com
Email: orders@arclereducation.com

ISBN: 978-1-77956-434-4 (Hardcover)

ABOUT THE AUTHOR

Carlos Andrés García is an environmental specialist with over 14 years of experience in climate change, sustainability, and international cooperation. He has led and managed projects aimed at fostering resilience, poverty reduction, and environmental sustainability. Currently serving as a Climate Change Consultant, Carlos has designed and supervised numerous climate change adaptation projects, securing international funding from bodies such as the Nordic Fund, USAID, and the Green Climate Fund. Throughout his career, Carlos has written numerous scientific articles focused on conservation programs, climate development, and environmental governance. His publications have contributed significantly to advancing knowledge in areas such as climate resilience, marine conservation, and environmental governance. Carlos's professional journey includes his work as a senior consultant for international development banks, where he assessed climate-resilient solutions for sectors like ports, coastal infrastructure, and agriculture. His expertise extends to navigating international environmental standards, promoting Nature-Based Solutions (NbS), and managing large-scale conservation programs. He has played a pivotal role in enhancing the environmental governance of protected areas and supporting local communities in adaptation strategies. In addition to his work in project management, Carlos holds a Master's in Environmental Economics from the University of Western Brittany in France and a Bachelor's in Business Administration from Colombia. He is also a certified Climate Change Officer. His dedication to sustainable development, along with his extensive writing and research experience, makes him a key contributor to climate action efforts across Latin America and globally.

TABLE OF CONTENTS

LIST OF FIGURES

LIST OF TABLES

PREFACE

Water resource management and flood control systems are vital components of sustainable development, essential for ensuring the efficient utilization of water resources while mitigating the adverse impacts of floods on communities and the environment. This book provides a comprehensive overview of the principles, techniques, and practices essential for effective water management and flood control.

This book comprises eight chapters; Chapter 1 introduces the foundational aspects of water resource management (WRM) and flood control. It introduces the challenges inherent in WRM, explains the significance of water resource conservation, and understands floods, their types, causes, and the importance of flood control, and forms. Chapter 2 navigates through the diverse types of water resources, from natural sources like surface water and glaciers to artificial sources such as wastewater reuse and groundwater, a vital yet often overlooked resource that receives special attention, along with techniques for its replenishment and storage.

Chapter 3 lays the groundwork for understanding the movement and distribution of water within the Earth's hydrologic cycle. Precipitation, evaporation, percolation, and groundwater dynamics are discussed. Chapter 4 focuses on flood risk assessment and management, essential for safeguarding lives and property from the devastating impacts of floods. It defines key concepts related to flood risk, delineates the tasks and components of flood risk management, and addresses the challenges inherent in this field.

In Chapter 5, flood control measures and management strategies take center stage. Structural and nonstructural methods for flood control are explored, alongside techniques for estimating flood damage and assessing net benefits. Chapter 6 focuses on flood forecasting, warning systems, and flood-fighting techniques. From forecasting procedures and runoff determination to engineering measures for flood fighting, this chapter provides insights into proactive strategies for managing flood events.

Chapter 7 delves into the integrated approach of water resource management (IWRM), emphasizing the need for holistic approaches to water management. The economic dimension of water, water management as a public good, and groundwater management are among the topics covered, highlighting the interconnectedness of water resources and socio-economic factors.

Finally, Chapter 8 explores the impact of climate change on water resources and the application of Geographic Information Systems (GIS) in water resource management and flood control. Understanding the ramifications of climate change and harnessing the power of GIS technology is essential for adapting to evolving environmental challenges.

This book has been written specifically for students and scholars to meet their needs in terms of knowledge and to provide them with a broad understanding of business information systems.

—Author

CHAPTER 1

INTRODUCTION TO WATER RESOURCE MANAGEMENT AND FLOOD

LEARNING OBJECTIVES

At the end of this chapter, readers will be able to:

1. Discuss the importance of sustainable water management in the context of limited freshwater resources.
2. Explore the significance of water resource conservation and management.
3. Describe techniques of water resource management, including top-down and bottom-up approaches.
4. Examine the historical context and evolution of water management practices.
5. Define integrated water resource management (IWRM) and its principles.
6. Explore the concept of managing water in urban settings, focusing on integrated urban water management (IUWM).
7. Understand the definition and types of floods, including fluvial, pluvial, and coastal.
8. Identify the causes of floods, such as heavy rainfall, snowmelt, and storm surges.

INTRODUCTORY EXAMPLE

Nestlé, a leading global food and beverage company, exemplifies the importance of water resource management in corporate sustainability efforts. With operations spanning over 190 countries, Nestlé recognizes the critical role of water in its business operations and the communities it serves.

Nestlé has implemented a comprehensive water stewardship strategy focused on responsible water use, conservation, and community engagement. For instance, in water-stressed regions like California, Nestlé has invested in water-saving technologies and practices in its factories to minimize water usage and optimize efficiency.

Moreover, Nestlé collaborates with local stakeholders and conservation organizations to support watershed restoration projects and promote sustainable water management practices. Through initiatives such as the Nestlé Waters Global Youth Summit and the Alliance for Water Stewardship, Nestlé engages with youth and communities to raise awareness about water conservation and empower individuals to become advocates for sustainable water management.

By prioritizing water stewardship across its value chain, Nestlé not only ensures the long-term viability of its business but also contributes to the preservation of water resources and the well-being of communities worldwide. This example highlights the proactive approach of companies like Nestlé in addressing water-related challenges and underscores the importance of collective action in achieving water security and resilience.

UNIT INTRODUCTION

Water resource management can sometimes be challenging as authors may have different interpretations of the term. Does it encompass all elements of the hydrological process or is it primarily focused on the aspects that directly impact humans, such as water consumption? Once the concept of "resource" is introduced, it inherently involves a human aspect (Gain et al., 2017). Water is an essential resource that we rely on, and there are effective methods to control its distribution. As a result, the concept of managing water resources is a practical and significant consideration.

Efficient and equitable use of water resources is crucial for sustainable growth. WRM (Water resource management) tackles a wide range of challenges, strategies, and practices to achieve this goal. Within this chapter, a thorough investigation is done on water resource management, starting with an analysis of the obstacles it faces in the present day. Comprehending the various challenges associated with water resources, such as reducing supplies, increasing demand, pollution, climate change effects, and governance complexities, is crucial for developing efficient solutions that preserve the health of the world's water ecosystems and cater to the requirements of both current and future generations (Fang et al., 2013).

An extensive examination of (WRM) water resource management is done in this chapter, a vital field that plays a crucial role in preserving ecosystems, bolstering livelihoods, and safeguarding the welfare of societies across the globe. Begin by discussing the current challenges in water resource management, and then delve into its importance in terms of economic prosperity, environmental sustainability, and social equity. At the center of this subject at hand are the various techniques and procedures used in water resource management, which include both top-down government actions and community-driven efforts (Benson & Lorenzoni, 2017). In addition, the historical background and development of water-related methods, are followed by an in-depth examination of flood dynamics and the significance of implementing flood control measures. During this exploration, the goal is to understand the intricacies of water resource management and its crucial role in creating strong communities and protecting water resources for generations to come (DiFrancesco & Tullos, 2014).

1.1. UNDERSTANDING WATER RESOURCE MANAGEMENT

Learning Objectives:

- Understand the challenges and implications of human activities on water resources.
- Recognize the significance of water conservation techniques for sustainability.
- Understand the concept of top-down and bottom-up approaches in water management.

Water resource management involves strategic planning, efficient distribution, and effective utilization of water resources. It falls under the general category of managing the water cycle. Water resource management at a city or district scale is like hydrology but on a smaller management scale compared to hydrology that includes the States or country. Water resource management focuses on implementing strategies to regulate water flow, considering factors like discharge from rivers and sediment movement (Morgan, 2014). Efficient water management plays a crucial role in preserving and safeguarding our precious water sources, which are scarce on the planet. So, basically, what exactly are water resources? These sources of water play a crucial role in supporting human life as they are essential for our existence. Almost every human activity relies on fresh water. Thus, water resource management focuses on optimizing the utilization of water resources and reducing losses in reservoirs and distribution networks.

Managing the distribution of water to different groups, and prioritizing needs like drinking, usage for industry, and agriculture, are all important aspects of water resource management (Ram & Irfan, 2021). There is a growing awareness of the impact of climate change on water resources, leading to an increased focus on improving the accuracy of water usage at different stages. Water resource management will require implementing advanced management strategies for multiple watersheds or interconnected systems. Internationally, water resource management is addressed through a comprehensive approach that involves the development and management of water, land, and other interconnected resources, as outlined by the Global Water Partnership. The goal is to optimize economic and social well-being while ensuring the long-term health of important ecosystems in a fair and balanced way (Ciampittiello et al., 2021). Now, delving into the topic of water resource management by addressing three crucial aspects which are discussed in subsections.

1.1.1. The Challenges of WRM

There are various aspects to consider when addressing this question, based on particular facts. The natural processes of surface run-off, precipitation, and groundwater levels ensure the sustainability of water resources. One of the most significant obstacles in the field of water resource management is the impact of human activities. Human activity has led to the pollution of freshwater resources such as rivers and lakes (Sheffield et al., 2018). Without proper implementation of sustainable water management, the future could potentially see conflicts arising over access to fresh water. Water is essential for sustaining life and is a fundamental requirement for the existence of all living beings on our planet. Many regions across the globe are currently facing severe water shortages due to prolonged droughts, which are believed to be a result of global warming.

There is a concerning issue with the state of many rivers, as they are experiencing depletion, contamination, and encroachment. Nowadays, the pressing concern revolves around the availability of freshwater for generations to come to sustain their livelihoods (DeBell et al., 2015). Freshwater is essential for various human needs such as livestock, sanitation, watering, drinking, and irrigation. There are options that humanity must take regarding their involvement in water resource management. Individuals need to consider the impact of their choices on water resource management to contribute to sustainable solutions. For instance, water-intensive crops with low nutritional value may face challenges in accessing irrigation, resulting in suboptimal growth. As a result, consumers may need to make concessions and consume smaller quantities (Elabras et al., 2013). Efforts to preserve water involve the transportation of water in rivers or pipes, which can have unintended effects on the groundwater, surface aquifers, and the surrounding ecosystem. Finding a solution for the issues faced by humans poses a significant challenge for the environment. The actions and behaviors of humans continue to complicate the lives of future generations, putting nature at risk. It all begins with the choices made regarding water consumption. Water resource management encompasses various aspects, including the interests of manufacturing and political goals (Ghobadi & Kang, 2023).

1.1.2. Significance of Water Resource Conservation and Management

The changing climate is causing shifts in water availability, highlighting the growing significance of water resource management. Efficient water management allows for the transfer of water from regions with surplus to those experiencing drought if there is a well-established distribution network connecting these areas. The scarcity of freshwater resources is a pressing issue, and implementing more efficient local water usage, especially in agriculture, can lead to significant water conservation (Bagatin et al., 2014). Water usage for controlling dust often gets disregarded, leading to strain on its availability in other crucial areas, particularly in regions near mining operations. Efficient water usage can be achieved by utilizing dust control items that effectively minimize or replace the quantity of water required for dust control. These products also ensure long-lasting performance in areas that need dust suppression. There are certain regions where wind traps are utilized to collect moisture from the surrounding air as a method of water harvesting (Huang et al., 2016).

The significance lies in the replenishment of reserves, which enhances the water condition in the regions where it happens. Given the ample supply of ocean water, the desalinization of seawater presents itself as a viable option to enhance water resource management. There is room for improvement in the efficiency of the current process, but the field of desalination shows promising potential. Repurposing gray water as an alternative to drinking water could be a valuable contribution to water conservation initiatives. The revival of local rainfall water capture and storage is crucial for conservation and water resource management. Given the ongoing disruptions caused by climate change, it is crucial to prioritize the preservation and oversight of water resources for long-term survival and sustainability (Bhatt et al., 2013).

1.1.3. Techniques of Water Resource Management

Water resource management often employs an integrated strategy to effectively implement optimal options for addressing water scarcity issues. There are two overarching methodologies for the process of organizing and overseeing tasks and resources. One method is hierarchical, with commands and control coming from the top. The other approach is decentralized, with initiatives and actions originating from the grassroots level (Hajkowicz & Higgins, 2008). Allowing to provide a more comprehensive explanation of the various elements of these strategies.

1.1.3.1. Top-Down Planning and Management

For many years, experts in water resource management have been dedicated to creating broad development strategies for various river basins across the globe. These plans usually include a set of reports, along with various appendices, which detail every aspect of managing and utilizing water resources. These documents analyze and assess several options for managing structures and non-structures. According to these assessments, the recommended course of action is the preferred plan (Koontz & Newig, 2014).

This strategic planning exercise has traditionally followed a top-down approach. Experts have been primarily in control of the strategy from the top down. This method often lacks active participation from interested parties. The approach presumes that there is a governing body with the necessary expertise and power to create and execute a plan. This body would be responsible for supervising and controlling the integrated growth and functioning of the activities that affect both groundwater and surface water within the basin. In today's changing landscape, there is a growing demand for reduced oversight by the government and increased public participation in management and planning activities. This shift in perspective has made strictly top-down strategies less desirable or suitable (Pissourios, 2014).

1.1.3.2. Bottom-Up Planning and Management

Within recent years, there has been a growing trend toward involving stakeholders in water resource management and planning. This approach ensures that those who may be impacted by decisions have a say in the procedure. Plans are being developed collaboratively, with input from all stakeholders, to ensure a consensus-based approach (Bossert et al., 1991). Collaboration between concerned citizens, nongovernmental organizations, and professionals in government agencies is on the rise as they strive to develop inclusive and flexible water management programs, rules, and strategies. Effective planning and management require the active involvement and engagement of all relevant stakeholders and advertisers in the process of water resource planning and management. This will require establishing a shared understanding of objectives and goals, as well as determining the most effective means to accomplish them. It would be ideal for this to happen before addressing any conflicting matters, allowing all parties involved to become acquainted and collaborate more efficiently (Gaymer et al., 2014) (Figure 1.1).

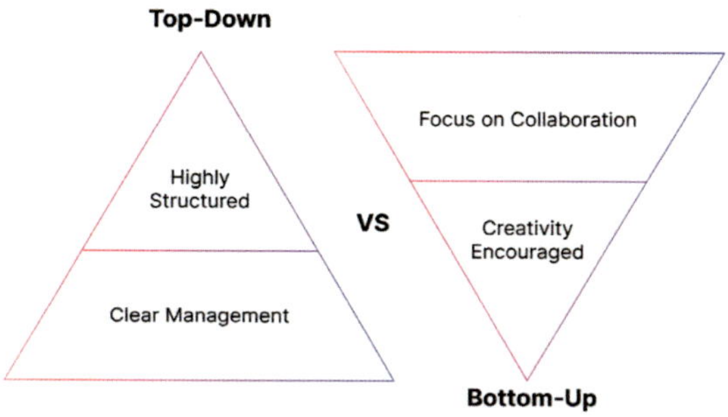

Figure 1.1. *Illustration of top-down and bottom-up planning.*

Source: Oliver, S.; Creative Commons License.

1.2. HISTORICAL CONTEXT AND EVOLUTION OF WATER MANAGEMENT

Learning Objectives:

- Understand the principles and development of Integrated Water Resource Management (IWRM).
- Recognize the Dublin Principles and their importance in sustainable water management.
- Explore the concept and benefits of Integrated Urban Water Management (IUWM).
- Understand the concept and goals of Integrated Urban Water Management (IUWM).

The history and development of water management has been a captivating journey using the annals of modern civilization, characterized by creativity, progress, and an unwavering commitment to long-term viability. Throughout history, the careful management of water assets has played a crucial role in shaping the development of societies, economies, and ecological systems, from ancient civilizations to present-day cities (Brown et al., 2009).

Throughout history, ancient civilizations including Sumerians and Egyptians ingeniously engineered irrigation systems to utilize the potential of rivers including the Nile, Tigris, and Euphrates. The early practices of managing water set the stage for agricultural success, allowing communities to thrive in areas with otherwise unpredictable weather patterns. The building of canals, dams, or reservoirs not only facilitated growing crops but also represented significant achievements in terms of societal progress and influence (Woodhouse & Muller, 2017).

The Roman Empire greatly improved methods for managing water through the development of aqueducts, which efficiently transported water across long distances to meet the needs of growing cities and populations. The ancient Romans were highly skilled in the field of hydraulic engineering. They were able to effectively and sustainably distribute water using underground conduits, aqueducts, and reservoirs. The impact of Roman water management can still be seen today, as numerous ancient structures continue to be utilized or serve as inspiration for contemporary infrastructure (Bjornlund & Bjornlund, 2019).

Did You Know?

The ancient town of Rome boasted a remarkable network of aqueducts, with a few still functioning to this day. The aqueducts were a testament to the impressive engineering skills of the Romans, as they efficiently transported fresh water from faraway sources to urban areas.

In the Middle Ages, the management of water became increasingly important as feudal societies faced the task of sustaining productivity in agriculture and protecting themselves from water-related disasters like droughts and floods. Monastic orders were instrumental in preserving and spreading hydraulic knowledge, improving methods for redirecting water, managing drainage, and mitigating floods. The rise of water mills or water-powered machinery highlighted the significance of water as a valuable energy source and catalyst for economic growth (Rahaman & Varis, 2005).

The Industrial Revolution brought about a significant change in the way water was managed, as the rapid growth of cities and industries created new challenges for water resources. The expansion of steam-powered machines and the growth in manufacturing industries required the development of extensive water supplies and treatment facilities for sewage. Urban planners and engineers developed creative solutions like centered water distribution networks, sewage systems, and storage facilities to address the increasing demands of expanding cities (Nicot et al., 2014). Water management in today's world has become more intricate and diverse. The increasing population, changing climate patterns, pollution, and degradation of ecosystems have put immense pressure on water resources, demanding comprehensive strategies for conservation and long-term sustainability. Collaboration between governments, international organizations, and grassroots movements, has begun to create comprehensive water management strategies that address the needs of different stakeholders, protect the environment, and promote social fairness (Benson et al., 2015).

The development of water management demonstrates humanity's continuous pursuit to control the immense potential of water while facing the unpredictable challenges of nature. Navigating through the challenges of the modern era, it is crucial to remember the valuable lessons from the past. These lessons emphasize the significance of being proactive, working together, and responsibly managing this invaluable asset to ensure its availability for future generations.

1.2.1. Integrated Water Resource Management

The concept of integrated water resources management (IWRM) involves coordinated growth and oversight of water, land, and associated resources. Its goal is to maximize social and economic well-being, while also preserving vital ecosystems.

According to some experts, there is a belief that IWRM and water security go hand in hand. Water security is seen as the ultimate objective, while IWRM is viewed as the essential process to reach that objective. The concept of Integrated Water Resource Management (IWRM) gained prominence during international conferences held in the latter 1900s and the beginning of the 2000s. However, it is important to note that collaborative water management organizations have been in existence for centuries. Conversations surrounding a comprehensive approach to water resource management emerged as early as the 1950s, culminating in the 1977 UN Water Conference (Mitchell, 2005).

The implementation of Integrated Water Resource Management (IWRM) was strongly advocated by the ministers within the final statement of the International Conference on Water or the Environment, held in 1992, commonly referred to as the Dublin Statement. This concept seeks to encourage alterations in practices that are seen as essential for enhancing water resource management. The topic of IWRM was discussed at the following World Water Forum, which had a more diverse group of stakeholders compared to previous conferences. This forum played a role in the establishment of the GWP (Saravanan et al., 2009).

The definition of IWRM provided by the International Water Organization is based on three principles that serve as the overall structure:

1. **Social Equity:** Ensuring that every individual, especially those who are marginalized and economically disadvantaged, has fair and unrestricted access to enough clean water that is essential for their overall well-being.
2. **Economic Efficiency:** Ensuring maximum benefit for a wide range of users using the resources at hand (Davis, 2007).
3. **Ecological Sustainability:** It is essential to recognize the importance of aquatic ecosystems and ensure that they receive sufficient resources to maintain their natural working.

The World Summit on Sustainable Development held in 2002, organized in Johannesburg addressed the development of IWRM, to promote its global implementation. The third World Water Forum emphasized the importance of Integrated Water Resource Management (IWRM) and highlighted the need for effective information sharing, active participation of stakeholders, and consideration of gender and class behavior.

Practically, IWRM approaches require the integration of knowledge from different fields and the input of various stakeholders to create and execute effective, fair, and lasting responses to water development and growth challenges. Therefore, IWRM serves as a holistic and inclusive approach to strategically plan and execute the management and enhancement of water resources. It aims to strike a harmonious equilibrium between societal and economic requirements while safeguarding the integrity of ecosystems for the benefit of future generations (McDonnell, 2008).

Furthermore, to support the accomplishment of Sustainable Development Goals (SDGs), the field of water resource management has been transforming into a more sustainable strategy. This involves adopting the Nexus approach, which considers the interaction of various sectors in managing water resources. The Nexus approach acknowledges the interconnectedness of water, energy, and food via global and local water, carbon, and energy cycles or chains.

An integrated approach to managing water resources focuses on preventing a fragmented approach by considering key aspects such as creating an enabling environment, defining roles for institutions, and implementing effective management instruments. It is crucial to consider various factors when implementing Integrated Water Resource Management (IWRM). These include the presence of political determination and commitment, the need for capacity development, sufficient sustainable cost recovery, investment, financial stability, as well as monitoring and evaluation (Biswas, 2004).

There isn't a single definitive administrative model. Mastering the science of Integrated Water Resource Management involves the skillful selection, adaptation, and application of the most suitable tools for any given scenario. The implementation of Integrated Water Resource Management practices relies heavily on the specific context. At the practical level, the main obstacle lies in effectively translating the agreed-upon principles into tangible actions (Ludwig et al., 2014) (Figure 1.2).

Figure 1.2. *Illustration of the aspect of integrated water resource management.*

Source: Paul Nixon; Creative Commons License.

1.2.2. Sustainable Water Management

Currently, a minuscule fraction of the world's clean water is readily available. Furthermore, there is a growing need for water in various sectors such as leisure, drinking, manufacturing, and agriculture. Managing the limited supply of fresh water from renewable sources has become an increasingly difficult task globally (Rao & Mamatha, 2004). A significant focus within water resource management involves maximizing water usage efficiency while minimizing its ecological footprint on the environment. Recognizing the importance of water as a crucial component of the ecosystem, integrated water resources management has emerged, drawing from the principles established in the 1992 Dublin conference (Haasnoot et al., 2011). Effective water management necessitates a comprehensive strategy rooted in Integrated Water Resource Management, which was initially outlined in 1992 during the Dublin (January) and Rio (July) seminars. Each of the four Dublin Rules, outlined in the Dublin Statement, are:

- Water is a limited and delicate resource that is crucial for supporting life, progress, and the natural world.
- Participatory methods are crucial for the effective development of water and management. It is important to involve users, policymakers, and planners, on every level to ensure success.
- Women are crucial in ensuring the provision, management, and protection of water resources.
- Water possesses economic significance in all of its applications and should be acknowledged as a valuable economic resource (Wilderer, 2004).
- Since 1992, the implementation of those principles has been instrumental in shaping the updating of the laws of national water management globally.

Additional obstacles to achieving sustainable and fair management of water resources arise from the presence of numerous reservoirs that are shared over various boundaries, whether they be international (as seen in water conflicts) or within a single nation (as observed in the Murray-Darling basin) (Pahl et al., 2008).

1.2.3. Managing Water in Urban Settings

IUWM Integrated urban water management to the systematic management of freshwater, wastewater, and stormwater within a specific geographic area, to develop a comprehensive basin-wide management strategy. It expands upon the previous issues for sanitation and access to water in an urban area by including the management of urban water throughout the whole river basin. To accomplish the objectives of Water Sensitive Urban Design IUWM is regarded as a strategic approach. The goal of IUWM is to transform the effects of growth in cities on the water cycle in nature (Farrelly & Brown, 2011). By approaching the management of urban water cycles holistically, one can achieve more effective utilization of resources, leading to economic advantages as well as enhanced social and ecological results. Implementing reuse strategies can help to develop an inner, urban, water cycle loop. To successfully create this water cycle loop, it is crucial to have a comprehensive grasp of each of the water balances before any development takes place and the water balance after development. Considering the flow of water in both the before and afterward development systems is crucial to minimizing the negative effects of urbanization on the natural cycle of water (Brown & Farrelly, 2009) (Figure 1.3).

Did You Know?

Several contemporary cities are adopting inventive strategies like green roofs and permeable pavements to effectively handle storm water runoff and mitigate the potential for urban flooding. These environmentally friendly practices not only assist in reducing the impact of floods but also improve the urban environment.

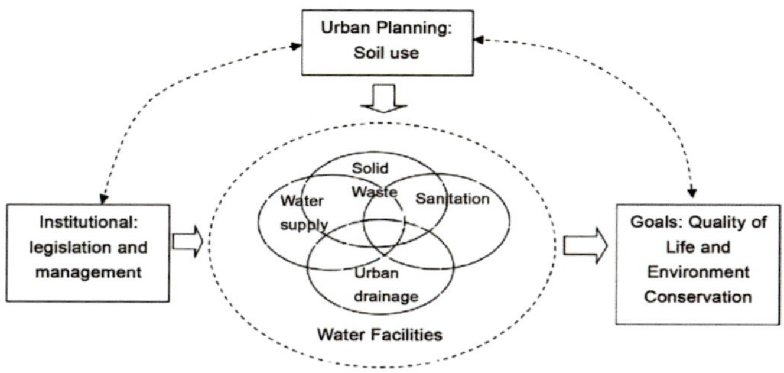

Figure 1.3. *Illustration of the components of integrated urban water management (IUWM).*

Source: Fernando Rosado; Creative Commons License.

Assessing the performance of new approaches to intervention from an urban water supply system can be done through a holistic approach that considers multiple system elements or criteria, including sustainability. This approach involves integrating water system parts such as stormwater subsystems, and wastewater, which can provide several advantages. Examining the flow of metabolism in urban water systems can provide valuable insights into the processes involved in the water cycle in cities of IUWM (Adem Esmail & Suleiman, 2020).

1.3. UNDERSTANDING FLOODS

Learning Objectives:

- Understand the types of flooding: fluvial, pluvial, and coastal.
- Explore the causes of flooding: heavy rainfall, king tides, storm surge.
- Know the factors influencing flood severity terrain, windstorm strength.
- Understand flood management strategies: dam releases, storm surge mitigation.

1.3.1. Definition of Flood

A flood occurs when water exceeds the typical boundaries of a watercourse. Whenever water covers land that is typically dry, flooding can occur. This scenario can occur when it is released from a natural waterway, like a river, lake, or creek. It can also occur when water is discharged from a reservoir, canal, or dam (Solin & Skubincan, 2013).

1.3.2. Types of Floods

There are three distinct categories of flooding: fluvial, flash, and coastal.

1.3.2.1. Fluvial Floods (River Floods)

Fluvial floods happen whenever the water level within a river, lake, or stream increases and spills over onto the surrounding land. The increase in the water level of a river may be a result of heavy rainfall or melting snow. River floods can have far-reaching consequences, impacting smaller rivers downstream and potentially leading to the failure of dams and dikes, resulting in the inundation of surrounding areas (Kundzewicz & Pińskwar, 2022) (Figure 1.4).

To determine the probability of river flooding, models consider past precipitation, forecasted precipitation, current river levels, as well as soil and terrain conditions. The degree of damage of a river's flood depends on factors such as the topography of the land, the length of time it rains, and the amount of rain in the surrounding area. Additional factors to consider are the saturation of soil water along with the impact of climate change on the amount and duration of rainfall. In low-lying regions, floodwater typically rises at a slower pace and is not as deep, yet it has the potential to linger for an extended period. In mountainous or hilly areas, flooding can occur rapidly following intense rainfall, recede swiftly, and result in significant damage caused by debris flow (Wheater, 2002).

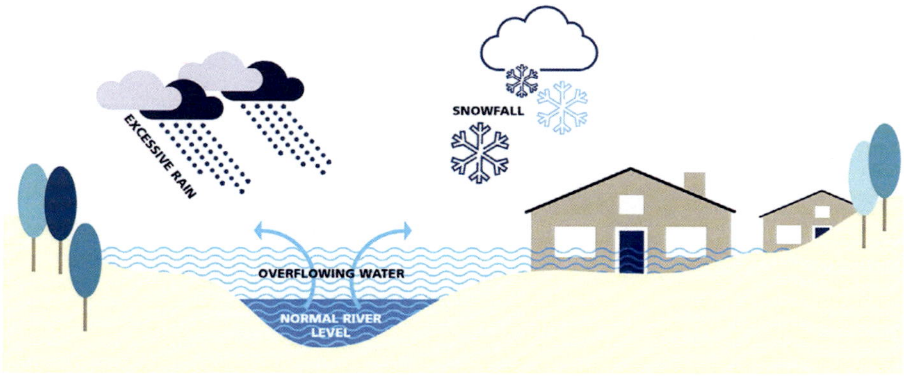

Figure 1.4. *Illustration of fluvial flood.*

Source: Zulkarnain Hassan; Creative Commons License.

1.3.2.2. Pluvial Floods (Flash Flood)

Pluvial floods occur when intense rainfall causes flooding without the involvement of overflowing water bodies. There is a common misconception surrounding floods, which is that being located close to a reservoir of water is the only way to be in danger. Pluvial flooding can occur in any location, whether it's urban or rural, regardless of the proximity to bodies of water (Acosta et al., 2018).

Pluvial flooding encompasses two typical kinds of flooding (Figure 1.5):

- Surface-level floods can occur when the capacity of a city's drainage system exceeds its limits, causing water to overflow onto streets and nearby buildings. The process happens gradually, allowing individuals to relocate to secure areas, and the water level is typically shallow (seldom exceeding one meter in depth). While there is no immediate danger to human lives, the potential for substantial economic harm exists (Zanchetta & Coulibaly, 2020).
- Flash floods are known for their powerful and rapid flow of water. These events are caused by heavy rainfall occurring rapidly in the area or on nearby higher ground. Flash floods may also happen due to a sudden discharge of water through an upstream levee along with a dam. Floods can pose significant risks and cause extensive damage, not only due to the powerful force of water but also the potential danger of debris carried along in the current (Hofmann & Schüttrumpf, 2019).

Figure 1.5. *Illustration of pluvial flood.*

Source: Zulkarnain Hassan, Creative Commons License.

1.3.2.3. Coastal Flood

Coastal flooding occurs when land across the coast is submerged by seawater. Storm surges are often the result of powerful windstorms, especially when get align with high tide. On the other hand, tsunamis can also lead to coastal flooding (Vousdoukas et al., 2016).

Storm surge is a common occurrence during windstorms, particularly during hurricanes or typhoons, where high winds push water onto the shore. It poses a significant danger in such situations. The outcome varies based on the tide. During periods of high tides, windstorms can lead to storm surge flooding that can have catastrophic consequences, causing extensive damage to property as well as lives in low-lying areas (Figure 1.6).

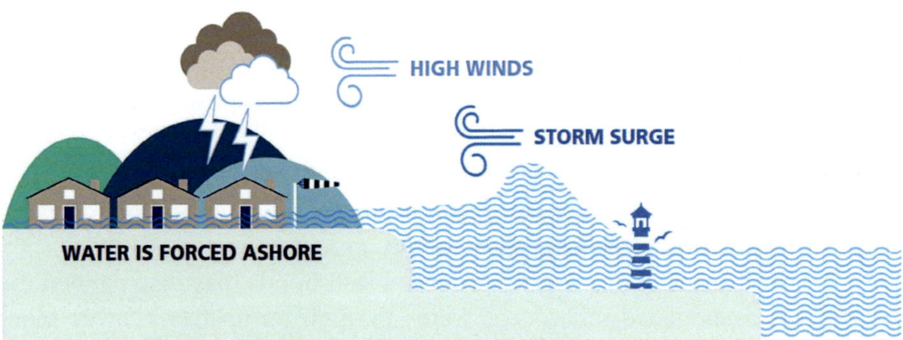

Figure 1.6. *Illustration of coastal flood.*

Source: Zulkarnain Hassan; Creative Commons License.

Several other factors contribute to the seriousness of a coastal flood, such as the intensity, size, velocity, and path of the windstorm. The topography of both land and water is also a crucial factor to consider. Coastal flood models consider various factors, including historical storm facts, to assess the likelihood and intensity of a storm surge (Woodruff et al., 2013).

1.3.3. Causes of Floods

Heavy rainfall or the quick melting of ice and snow can result in flooding. Geography can contribute to increased flood risk in certain regions (Merz et al., 2021). A few significant factors that contribute to flooding are discussed in subsections.

1.3.3.1. Heavy Rainfall

Heavy rainfall is the primary factor behind riverine and sudden flooding. Rivers are created through a gradual process that spans thousands of years. Each river has its unique characteristics and is shaped by various factors:

- The typical amount of precipitation and runoff in the area;
- The characteristics of the local area include vegetation, topography, and soil types.

The characteristics of these features remain consistent over time, except for changes in rainfall. Rivers have a limit to the amount of water being held. Increased precipitation results in a higher volume of water runoff. The excess runoff overflows on the surrounding land due to the river channel's inability to accommodate it. When there is a significant time gap between rainfall and flooding exceeding six hours, a riverine flood occurs (Chen et al., 2007).

Practice Problem

In a region covering 500 square kilometers, with 60% of its area comprised of impermeable surfaces, heavy rainfall poses a significant flood risk. With an average annual rainfall of 1200 millimeters, a single intense rainfall event of 150 millimeters over 24 hours can lead to substantial runoff. To assess the potential volume of runoff generated, we need to calculate the total rainfall and then determine the volume of runoff considering the impermeable surfaces. Utilizing the Soil Conservation Service (SCS) Runoff Curve Number (CN) method, assuming a CN value of 75 for the region, we can estimate the runoff coefficient and subsequently calculate the runoff volume. Such calculations are critical for understanding flood risks associated with heavy rainfall events and inform the implementation of appropriate water resource management and flood control measures.

Solutions to Practice Problem

Considering the total area of 500 square kilometers and an average annual rainfall of 1200 millimeters, heavy rainfall of 150 millimeters over 24 hours results in substantial total rainfall. By factoring in the impermeable surfaces covering 60% of the region, we estimate the permeable area available for infiltration. Applying the SCS Runoff Curve Number method, with a chosen CN value of 75, we derive the runoff coefficient and use it to calculate the volume of runoff. This estimation provides insight into the potential flood risk posed by heavy rainfall events, highlighting the importance of effective water resource management strategies and flood control measures to mitigate adverse impacts on communities and infrastructure within the region.

1.3.3.2. King Tide

An exceptionally large tide is commonly referred to as a 'king tide.' The fluctuating levels in these tides are an inevitable process that follows a predictable cycle. The timing of these events can vary depending on the specific location and from year to year. It can have significant impacts in areas where both land and water intersect, such as estuaries, harbors, beaches, and other coastal areas. King tides have the potential for increasing the effects and scope of riverine flooding (Chanson et al., 2012). Picture a coastal town situated next to a river that flows into the ocean. In the event of a river flooding, certain areas of the town could potentially be submerged. When a flood coincides with an elevated king tide, the floodwater has limited chances to drain into the sea. There is a higher probability that a larger portion of the town will experience flooding at a more severe level (Román & Ellis, 2018).

1.3.3.3. Storm Surge

Storm surge occurs when a storm causes the seawater levels along a shore to rise above its normal level. During the storm, there are usually powerful winds blowing towards the shore, which is caused by a decrease in the pressure of the air. Tropical cyclones are frequently accompanied by storm surges. Other than tropical regions, storm surges can occur because of a powerful low-pressure system. Coastal flooding may occur over a storm surge. Furthermore, when a storm surge coincides with a riverine flood, it can lead to an expansion within the area and magnitude of flooding (Bode & Hardy, 1997).

1.3.3.4. Snow Melt and Dam Releases

Snow is like a water reservoir, patiently waiting for the right conditions to flow downhill. When temperatures rise very quickly, snow can melt at a fast pace. When this occurs, it can result in the release of large amounts of water that exceed the capacity of the river channels downstream, leading to flooding in the affected areas (Kumar et al., 2007).

Each dam has a limit to the amount of water it can hold, regardless of its size. When a dam reaches its maximum capacity, operators may need to release water, which can potentially lead to flooding. Work is done together with dam operators during flood events to provide them with up-to-date weather information. Operators utilize this data to effectively manage their dams, ensuring minimal impacts on the communities located downstream (Arheimer et al., 2017).

Did You Know?

China experienced the world's largest flooding in recorded history in 1931 when the Yangtze and Huai streams overflowed over the summer monsoon months. The devastating floods had a profound impact on several million people, resulting in widespread destruction and underscoring the significant consequences that floods can bring to densely populated areas.

1.4. IMPORTANCE OF FLOOD CONTROL

Learning Objectives:

- Understand the importance of flood control for safeguarding lives and property.

Flood control is crucial for a variety of reasons, including safeguarding human life and property, preserving ecosystems, and promoting sustainable growth. Ensuring the safety of human lives is of utmost importance, and thus, implementing efficient flood control measures is crucial (Lechowska, 2022). Floods can present substantial dangers to families, individuals, and entire communities, resulting in injuries, relocation, and even loss of life. Flood control projects mitigate the human impact of natural disasters by employing tactics that include early warning systems, evacuation plans, and robust infrastructure (Ren et al., 2017).

Furthermore, flood control is essential for safeguarding expensive property and infrastructure. Floodwater has the potential to inflict significant harm on residences, commercial establishments, farmlands, and public infrastructure, including educational institutions and medical centers. Communities can mitigate the potential harm caused by flooding events by constructing flood control infrastructure such as levees, floodwalls, and stormwater management systems. This helps to minimize harm to property and economic losses (Haeuber & Michener, 1998).

Moreover, flood control is crucial for upholding economic stability and fostering sustainable growth. Floods have the potential to disturb transport systems, causing a halt in the movement of products and services and hindering economic activity. Flood control systems protect essential infrastructure such as bridges, roads, and utilities, ensuring that cities can operate effectively even during severe weather events. This resilience is essential for attracting investment, bolstering local firms, and cultivating sustainable long-term success (Rezende et al., 2019).

Floods can cause significant environmental repercussions, besides their effects on humans and the economy. Unregulated inundation can result in the erosion of soil, contaminating water, and loss of habitats, posing a threat to the well-being of habitats and the organisms that live inside them. Flood management measures frequently integrate ecological factors, such as the conservation of wetlands and natural floodplains. These areas act as natural barriers against floods and offer important wildlife habitats.

SUMMARY

- The chapter "Introduction to Water Resource Management and Flood" provides a comprehensive overview of water management principles and flood dynamics. It begins by exploring the complexities of water resource management (WRM), highlighting challenges such as population growth, urbanization, and climate change.
- Emphasizing the significance of conservation and sustainable management, the chapter delves into techniques including top-down governance and bottom-up community involvement. Moving through the historical evolution of water management, it introduces concepts like Integrated Water Resource Management (IWRM) and Sustainable Water Management, stressing their importance in addressing modern water challenges. Urban water management strategies are also discussed, focusing on the unique needs of cities.
- Transitioning to flood dynamics, the chapter defines floods and categorizes them into fluvial, pluvial, and coastal types, each with distinct causes and impacts. It elucidates factors contributing to floods, such as heavy rainfall, king tides, and storm surges, underscoring the importance of understanding these triggers for effective flood mitigation.
- Finally, the chapter underscores the criticality of flood control measures in safeguarding communities and ecosystems against the devastating consequences of inundation events. Overall, the chapter serves as a foundational guide to understanding the complexities of water resource management and the dynamics of flooding, offering insights into strategies for sustainable water management and flood resilience.

REVIEW QUESTIONS

1. What are the primary challenges faced in water resource management?
2. Why is water resource conservation and management significant for sustainable development?
3. What are some techniques used in water resource management?
4. How do top-down and bottom-up planning and management approaches differ?
5. How has water management evolved historically?

6. What is Integrated Water Resource Management (IWRM) and why is it important?
7. What distinguishes sustainable water management from conventional practices?
8. What are the unique challenges associated with managing water in urban settings?
9. Define different types of floods and their primary causes.

MULTIPLE CHOICE QUESTIONS

1. What are the challenges typically faced in Water Resource Management (WRM)?

a. Political instability
b. Decreasing population
c. Pollution
d. Abundant water supply

2. Why is Water Resource Conservation and Management significant?

a. It has no significant impact on society.
b. It ensures equitable access to water.
c. It increases pollution levels.
d. It leads to wasteful water usage.

3. Which of the following is an example of Top-Down Planning and Management in WRM?

a. Community-driven initiatives
b. Grassroots involvement
c. Governmental regulations and policies
d. Local knowledge integration

4. Integrated Water Resource Management (IWRM) emphasizes:

a. Fragmentation of water management practices
b. Isolation of different sectors involved in water management
c. Holistic and integrated approaches to water management
d. Disregarding environmental concerns

5. What is one of the primary causes of Pluvial Floods (Flash Flood)?

a. River overflow
b. Snow melt

c. Intense rainfall

d. Storm surge

6. Why is Flood Control important?

a. It has no significant impact on communities.

b. It helps mitigate the socio-economic impacts of flooding.

c. It exacerbates the effects of flooding.

d. It increases the frequency of flooding events.

Answers to Multiple Questions

1. (c); 2. (b); 3. (c); 4. (c); 5. (c); 6. (b)

REFERENCES

1. Acosta-Coll, M., Ballester-Merelo, F., Martinez-Peiró, M., & De la Hoz-Franco, E. (2018). Real-time early warning system design for pluvial flash floods – A review. *Sensors, 18*(7), 22–55. https://doi.org/10.3390/s18072055.

2. Adem Esmail, B., & Suleiman, L. (2020). Analyzing evidence of sustainable urban water management systems: A review through the lenses of sociotechnical transitions. *Sustainability, 12*(11), 44–81. https://doi.org/10.3390/su12114481.

3. Arheimer, B., Donnelly, C., & Lindström, G. (2017). Regulation of snow-fed rivers affects flow regimes more than climate change. *Nature Communications, 8*(1), 62–89. https://doi.org/10.1038/s41467-017-00062-w.

4. Bagatin, R., Klemeš, J. J., Reverberi, A. P., & Huisingh, D. (2014). Conservation and improvements in water resource management: A global challenge. *Journal of Cleaner Production, 77*, 1–9. https://doi.org/10.1016/j.jclepro.2014.04.021.

5. Benson, D., & Lorenzoni, I. (2017). Climate change adaptation, flood risks, and policy coherence in integrated water resources management in England. *Regional Environmental Change, 17*, 1921–1932. https://doi.org/10.1007/s10113-017-1153-4.

6. Benson, D., Gain, A. K., & Rouillard, J. J. (2015). Water governance in a comparative perspective: From IWRM to a 'nexus' approach? *Water Alternatives, 8*(1), 756–773. http://www.water-alternatives.org/index.php/alldoc/articles/vol8/v8issue1/299-a8-1-12/file (accessed on 27 June 2024).

7. Bhatt, R. A. J. A. N., Kukal, S. S., & Arora, S. (2013). Resource conservation technologies for improving water productivity. *Journal of Soil and Water Conservation, 12*(4), 313–320. https://doi.org/10.1016/j.still.2012.10.005.

8. Biswas, A. K. (2004). Integrated water resources management: A reassessment: A water forum contribution. *Water International, 29*(2), 248–256. https://doi.org/10.1080/02508060408691785.

9. Bjornlund, V., & Bjornlund, H. (2019). Understanding agricultural water management in a historical context using a socioeconomic and biophysical framework. *Agricultural Water Management, 213*, 454–467. https://doi.org/10.1016/j.agwat.2018.12.022.

10. Bode, L., & Hardy, T. A. (1997). Progress and recent developments in storm surge modeling. *Journal of Hydraulic Engineering, 123*(4), 315–331. https://doi.org/10.1061/(ASCE)0733-9429(1997)123:4(315).

11. Bossert, T., Soebekti, R., & Rai, N. K. (1991). 'Bottom-up' planning in Indonesia: Decentralization in the Ministry of Health. *Health Policy and Planning, 6*(1), 55–63. https://doi.org/10.1093/heapol/6.1.55.

12. Brown, R. R., & Farrelly, M. A. (2009). Delivering sustainable urban water management: A review of the hurdles we face. *Water Science and Technology, 59*(5), 839–846. https://doi.org/10.2166/wst.2009.103.

13. Brown, R. R., Keath, N., & Wong, T. H. (2009). Urban water management in cities: Historical, current and future regimes. *Water Science and Technology, 59*(5), 847–855. https://doi.org/10.2166/wst.2009.104.

14. Chanson, H., Brown, R., & Trevethan, M. (2012). Turbulence measurements in a small subtropical estuary under king tide conditions. *Environmental Fluid Mechanics, 12*, 265–289. https://doi.org/10.1007/s10652-011-9212-1.

15. Chen, C. S., Chen, Y. L., Liu, C. L., Lin, P. L., & Chen, W. C. (2007). Statistics of heavy rainfall occurrences in Taiwan. *Weather and Forecasting, 22*(5), 981–1002. https://doi.org/10.1175/WAF1033.1.

16. Ciampittiello, M., Dresti, C., & Saidi, H. (2021). Water resource management through an understanding of the water balance components: A case study of a Sub-Alpine Shallow Lake. *Water, 13*(21), 3124. https://doi.org/10.3390/w13213124.

17. Davis, M. D. (2007). Integrated water resource management and water sharing. *Journal of Water Resources Planning and Management, 133*(5), 427–445. https://doi.org/10.1061/(ASCE)0733-9496(2007)133:5(427).

18. DeBell, L., Anderson, K., Brazier, R. E., King, N., & Jones, L. (2015).

Water resource management at catchment scales using lightweight UAVs: Current capabilities and future perspectives. *Journal of Unmanned Vehicle Systems, 4*(1), 7–30. https://doi.org/10.1139/juvs-2015-0002.

19. DiFrancesco, K. N., & Tullos, D. D. (2014). Flexibility in water resources management: Review of concepts and development of assessment measures for flood management systems. *JAWRA Journal of the American Water Resources Association, 50*(6), 1527–1539. https://doi.org/10.1111/jawr.12222.
20. Elabras Veiga, L. B., & Magrini, A. (2013). The Brazilian water resources management policy: Fifteen years of success and challenges. *Water Resources Management, 27*, 2287–2302. https://doi.org/10.1007/s11269-013-0322-6.
21. Fang, S., Xu, L., Pei, H., Liu, Y., Liu, Z., Zhu, Y., & Zhang, H. (2013). An integrated approach to snowmelt flood forecasting in water resource management. *IEEE Transactions on Industrial Informatics, 10*(1), 548–558. https://doi.org/10.1109/TII.2013.2244193.
22. Farrelly, M., & Brown, R. (2011). Rethinking urban water management: Experimentation as a way forward? *Global Environmental Change, 21*(2), 721–732. https://doi.org/10.1016/j.gloenvcha.2011.02.007.
23. Gain, A. K., Mondal, M. S., & Rahman, R. (2017). From flood control to water management: A journey of Bangladesh towards integrated water resources management. *Water, 9*(1), 2–55. https://doi.org/10.3390/w9010002.
24. Gaymer, C. F., Stadel, A. V., Ban, N. C., Cárcamo, P. F., Ierna Jr, J., & Lieberknecht, L. M. (2014). Merging top-down and bottom-up approaches in marine protected areas planning: Experiences from around the globe. *Aquatic Conservation: Marine and Freshwater Ecosystems, 24*(S2), 128–144. https://doi.org/10.1002/aqc.2465.
25. Ghobadi, F., & Kang, D. (2023). Application of machine learning in water resources management: A systematic literature review. *Water, 15*(4), 620. https://doi.org/10.3390/w15040620.
26. Haasnoot, M., Middelkoop, H., Van Beek, E., & Van Deursen, W. P. A. (2011). A method to develop sustainable water management strategies for an uncertain future. *Sustainable Development, 19*(6), 369–381. https://doi.org/10.1002/sd.518.
27. Haeuber, R. A., & Michener, W. K. (1998). Natural flood control. *Issues in Science and Technology, 15*(1), 74–80. https://www.jstor.org/stable/43312677.

28. Hajkowicz, S., & Higgins, A. (2008). A comparison of multiple criteria analysis techniques for water resource management. *European Journal of Operational Research, 184*(1), 255–265. https://doi.org/10.1016/j.ejor.2006.10.031.
29. Hofmann, J., & Schüttrumpf, H. (2019). Risk-based early warning system for pluvial flash floods: Approaches and foundations. *Geosciences, 9*(3), 127. https://doi.org/10.3390/geosciences9030127.
30. Huang, C. L., Yu, C. P., Lin, T., & Ye, Z. (2016). Water conservation significance of municipal solid waste management: A case of Xiamen in China. *Journal of Cleaner Production, 129*, 693–703. https://doi.org/10.1016/j.jclepro.2016.04.075.
31. Koontz, T. M., & Newig, J. (2014). From planning to implementation: Top-down and bottom-up approaches for collaborative watershed management. *Policy Studies Journal, 42*(3), 416–442. https://doi.org/10.1111/psj.12068.
32. Kumar, V., Singh, P., & Singh, V. (2007). Snow and glacier melt contribution in the Beas River at Pandoh dam, Himachal Pradesh, India. *Hydrological Sciences Journal, 52*(2), 376–388. https://doi.org/10.1623/hysj.52.2.376.
33. Kundzewicz, Z. W., & Pińskwar, I. (2022). Are pluvial and fluvial floods on the rise? *Water, 14*(17), 2612. https://doi.org/10.3390/w14172612.
34. Lechowska, E. (2022). Approaches in research on flood risk perception and their importance in flood risk management: A review. *Natural Hazards, 111*(3), 2343–2378. https://doi.org/10.1007/s11069-021-04867-2.
35. Ludwig, F., van Slobbe, E., & Cofino, W. (2014). Climate change adaptation and integrated water resource management in the water sector. *Journal of Hydrology, 518*, 235–242. https://doi.org/10.1016/j.jhydrol.2014.01.010.
36. McDonnell, R. A. (2008). Challenges for integrated water resources management: How do we provide the knowledge to support truly integrated thinking? *International Journal of Water Resources Development, 24*(1), 131–143. https://doi.org/10.1080/07900620701723545.
37. Merz, B., Blöschl, G., Vorogushyn, S., Dottori, F., Aerts, J. C., Bates, P., & Macdonald, E. (2021). Causes, impacts, and patterns of disastrous river floods. *Nature Reviews Earth & Environment, 2*(9), 592–609. https://doi.org/10.1038/s43017-021-00206-y.
38. Mitchell, B. (2005). Integrated water resource management, institutional arrangements, and land-use planning. *Environment and Planning A, 37*(8), 1335–1352. https://doi.org/10.1068/a37374.
39. Morgan, E. (2014). Science in sustainability: A theoretical framework

for understanding the science-policy interface in sustainable water resource management. *The International Journal of Sustainability Policy and Practice, 9*(2), 37–54. https://doi.org/10.18848/2325-1166/CGP/v09i02/37366.

40. Nicot, J. P., Scanlon, B. R., Reedy, R. C., & Costley, R. A. (2014). Source and fate of hydraulic fracturing water in the Barnett Shale: A historical perspective. *Environmental Science & Technology, 48*(4), 2464–2471. https://doi.org/10.1021/es4054325.
41. Pahl-Wostl, C., Tàbara, D., Bouwen, R., Craps, M., Dewulf, A., Mostert, E., & Taillieu, T. (2008). The importance of social learning and culture for sustainable water management. *Ecological Economics, 64*(3), 484–495. https://doi.org/10.1016/j.ecolecon.2007.08.009.
42. Pissourios, I. (2014). Top-down and bottom-up urban and regional planning: Towards a framework for the use of planning standards. *European Spatial Research and Policy, 21*(1), 83–99. https://doi.org/10.2478/esrp-2014-0005.
43. Rahaman, M. M., & Varis, O. (2005). Integrated water resources management: Evolution, prospects, and future challenges. *Sustainability: Science, Practice and Policy, 1*(1), 15–21. https://doi.org/10.1080/15487733.2005.11907966.
44. Ram, S. A., & Irfan, Z. B. (2021). Application of System Thinking Causal Loop Modelling in understanding water Crisis in India: A case for sustainable Integrated Water resources management across sectors. *Hydro Research, 4*, 1–10. https://doi.org/10.2166/hyd.2021.050.
45. Rao, S. M., & Mamatha, P. (2004). Water quality in sustainable water management. *Current Science, 87,* 942–947. Retrieved from: https://www.currentscience.ac.in (accessed on 27 June 2024).
46. Ren, M., He, X., Kan, G., Wang, F., Zhang, H., Li, H., & Zhang, Z. (2017). A comparison of flood control standards for reservoir engineering for different countries. *Water, 9*(3), 152. https://doi.org/10.3390/w9030152.
47. Rezende, O. M., de Oliveira, A. K. B., Jacob, A. C. P., & Miguez, M. G. (2019). A framework to introduce urban flood resilience into the design of flood control alternatives. *Journal of Hydrology, 576*, 478–493. https://doi.org/10.1016/j.jhydrol.2019.05.052.
48. Román-Rivera, M. A., & Ellis, J. T. (2018). The king tide conundrum. *Journal of Coastal Research, 34*(4), 769–771. https://doi.org/10.2112/JCOASTRES-D-18-00057.1.
49. Saravanan, V. S., McDonald, G. T., & Mollinga, P. P. (2009). A critical review of integrated water resources management: Moving beyond

polarized discourse. *Natural Resources Forum, 33*(1), 76–86. https://doi.org/10.1111/j.1477-8947.2009.01208.x.

50. Sheffield, J., Wood, E. F., Pan, M., Beck, H., Coccia, G., Serrat-Capdevila, A., & Verbist, K. (2018). Satellite remote sensing for water resources management: Potential for supporting sustainable development in data-poor regions. *Water Resources Research, 54*(12), 9724–9758. https://doi.org/10.1029/2018WR022934.
51. Solin, L., & Skubincan, P. (2013). Flood risk assessment and management: Review of concepts, definitions, and methods. *Geogr. J, 65*, 23–44. https://doi.org/10.1111/geog.12017.
52. Vousdoukas, M. I., Voukouvalas, E., Mentaschi, L., Dottori, F., Giardino, A., Bouziotas, D., & Feyen, L. (2016). Developments in large-scale coastal flood hazard mapping. *Natural Hazards and Earth System Sciences, 16*(8), 1841–1853. https://doi.org/10.5194/nhess-16-1841-2016.
53. Wheater, H. S. (2002). Progress in and prospects for fluvial flood modeling. *Philosophical Transactions of the Royal Society of London. Series A: Mathematical, Physical and Engineering Sciences, 360*(1796), 1409–1431. https://doi.org/10.1098/rsta.2002.1001.
54. Wilderer, P. A. (2004). Applying sustainable water management concepts in rural and urban areas: Some thoughts about reasons, means and needs. *Water Science and Technology, 49*(7), 7–16. https://doi.org/10.2166/wst.2004.0607.
55. Woodhouse, P., & Muller, M. (2017). Water governance – A historical perspective on current debates. *World Development, 92*, 225–241. https://doi.org/10.1016/j.worlddev.2016.11.004.
56. Woodruff, J. D., Irish, J. L., & Camargo, S. J. (2013). Coastal flooding by tropical cyclones and sea-level rise. *Nature, 504*(7478), 44–52. https://doi.org/10.1038/nature12855.
57. Yevjevich, V. (1974). Systematization of flood control measures. *Journal of the Hydraulics Division, 100*(11), 1537–1548.
58. Zanchetta, A. D., & Coulibaly, P. (2020). Recent advances in real-time pluvial flash flood forecasting. *Water, 12*(2), 570. https://doi.org/10.3390/w12020570.

CHAPTER 2

TYPES OF WATER RESOURCES AND MANAGEMENT

LEARNING OBJECTIVES

At the end of this chapter, readers will be able to:

1. Identify natural sources of freshwater and their importance for humans.
2. Understand the significance of surface water resources for various activities.
3. Explore the role of glaciers in providing freshwater resources.
4. Know the concept of artificial sources of usable water.
5. Describe the processes and benefits of wastewater reuse.
6. Discuss the methods and challenges of desalinating seawater.
7. Recognize the importance of groundwater resources and their replenishment mechanisms.

INTRODUCTORY EXAMPLE

Coastal Beverage Corp, a beverage packaging company located along the coast, is facing significant water scarcity issues due to increasing demand and saltwater intrusion into freshwater sources. To address these challenges, Coastal Beverage Corp has implemented state-of-the-art water management strategies.

Firstly, the company has invested in desalination plants that convert seawater into usable water for their manufacturing operations, ensuring a reliable water supply even during severe droughts. Additionally, they have introduced a closed-loop system within their facilities to recycle and treat wastewater produced during production. This system minimizes discharge and conserves freshwater resources.

Furthermore, Coastal Beverage Corp collaborates with local communities on groundwater restoration initiatives. These initiatives involve replenishing aquifers by carefully infiltrating rainwater and surface water runoff. By integrating these approaches, Coastal Beverage Corp not only ensures a sustainable water supply for its operations but also contributes to strengthening the local water supply, showcasing effective water resource management within an industrial setting.

UNIT INTRODUCTION

Water is a crucial resource that supports life on Earth, performing a vital role in numerous aspects of human civilization, including agriculture, health, industry, and sanitation. Gaining knowledge about various categories of watersheds and management is essential to guarantee the long-term availability of clean and useful water for both present and future generations.

Throughout this chapter, a vast array of water resources is explored, including artificial and natural sources of freshwater. In addition, the complex processes that contribute to the replenishment, storage, and utilization of these resources are also observed (McKinney & Cai, 2002).

As exploring further toward the Earth's crust, the extensive underground water sources that exist beneath the surface are discovered, providing sustenance for both ecosystems and human populations. In addition, the processes that contribute to the replenishment of groundwater supplies are studied, such as the condensation of atmospheric humidity, the infiltration of surface water from streams and lakes, and rainfall (Snelder & Biggs, 2002).

Ultimately, the significance of storing groundwater is explored as a crucial safeguard against seasonal variations and droughts, emphasizing the importance of sustainable management procedures in preserving this invaluable resource for future generations.

2.1. NATURAL SOURCES OF FRESH WATER

Learning Objectives:

- Understand the distribution of Earth's water resources: freshwater vs. saltwater.
- Understand the natural sources of freshwater: surface water, groundwater, glaciers.

Water resources are valuable assets that provide essential support for human needs, such as drinking water supply and irrigation. Most of the Earth's water is composed of salt water, with only a small percentage being fresh water. A significant portion of this freshwater is locked out in glaciers and polar ice caps. Most of the freshwater that is not frozen is primarily located underground, while a small portion can be found on the Earth's surface or in the atmosphere. There are various sources of fresh water, such as groundwater, surface water, river flow, and frozen water. Various types of freshwaters are not naturally occurring, but rather created by humans. These sources might consist of treated wastewater which can be reused, as well as seawater which has been desalinated. Water resources are utilized by individuals for a variety of purposes, including household activities, agricultural, and industrial. Around 70% of the total surface of the earth is comprised of water. It is quite difficult to comprehend the sheer magnitude of water on the planet. The earth's total water resources are approximately 326 million cubic miles, and each cubic mile is equivalent to about 1 trillion gallons of water. Visualize the vastness of one trillion gallons of water by envisioning 40 million swimming pools or 24 billion baths (Epstein & Mayeda, 1953). Now, take those figures and multiply them by 326 million!

Out of the vast amount of water available, a mere 2.5% is freshwater, while the remaining 97.5% consists of saltwater. Approximately 69% of the Earth's freshwater resources are found in glaciers and ice caps, while groundwater accounts for around 30%. Surface water, on the other hand, makes up a very small percentage, only 0.27%. Although various water resources play a crucial role in sustaining the planet, the availability of freshwater holds particular significance for human beings (Nagrodski et al., 2012).

2.1.1. Surface Water Resources

Surface water refers to the presence of water in natural bodies such as rivers, lakes, or freshwater wetlands. Surface water is replenished by precipitation and lost via evapotranspiration, discharge to the oceans, evaporation, and groundwater recharge. Precipitation within the basin is the sole natural input for any surface water system. Several other factors also play a role in determining the total amount of water within the system at any given moment. Various factors play a role in water resource management, such as the capability of storage bodies like lakes, wetlands, and reservoirs, the permeability of the underlying soil, the runoff properties of the surrounding land, and the timing for precipitation and local evaporation rates. These factors also impact the percentage of water loss.

People frequently enhance storage capacity through the construction of reservoirs and reduce it through draining wetlands. Human activities can have a significant impact on the flow of water, particularly when areas are paved and stream channels are modified (Weatherhead & Howden, 2009).

Water serves a variety of purposes, including providing safe drinking water, supporting recreational activities, facilitating irrigation for crops, powering industries, sustaining livestock, enabling transportation, and generating hydroelectric energy. More than 63% of the water used for public consumption is sourced from surface water. Surface water is responsible for supplying 58% of the water used in irrigation. Most of the water used by industry is sourced via surface water systems (Naidoo & Olaniran, 2014). Thus, the preservation and enhancement of surface water conservation and quality are crucial.

2.1.2. Water From Glaciers

Glacier rainwater is also recognized as a natural source of water. The Himalayas, known as "The Roof of the World," boast vast and challenging high-altitude regions, along with the largest expanse of glaciers and permafrost beyond the Polar Regions. Asia is home to 10 of the largest rivers in the region, which play a vital role in supporting the livelihoods of over a billion people. Adding to the complexity, temperatures in that area are increasing at a faster rate than the worldwide average. In Nepal, the average annual temperature has increased by 0.6°C in the past decade, while worldwide, the Earth has experienced a warming of around 0.7°C in the last century (Shreve, 1972).

2.2. ARTIFICIAL SOURCES OF USABLE WATER

Learning Objectives:

- Understand artificial sources of fresh water: wastewater reuse, desalination.
- Recognize the various purposes and types of wastewater reuse.

Various sources of fresh water are not naturally occurring. These involve treated wastewater (reclaimed water), atmospheric water generators, and desalinated seawater. Nevertheless, it is crucial to consider the economic and environmental consequences of these technologies (Park et al., 2003).

2.2.1. Wastewater Reuse

Water reclamation, also known as water recycling or wastewater reuse involves transforming municipal or commercial wastewater into reusable water for various purposes. There are various types of water reuse, such as urban reuse, irrigation, planned potable reuse, industrial reuse, environmental reuse, agricultural reuse, and de facto wastewater reuse (unplanned potable reuse). As an illustration, reuse can involve activities such as irrigating gardens and agricultural fields or replenishing surface water and groundwater, also known as groundwater recharge. Water that has been recycled can also be used to meet various needs in homes (such as flushing toilets), businesses, and industries. It can even undergo treatment to meet the standards for drinking water (Adewumi et al., 2010). Direct potable reuse refers to the process of injecting reclaimed water through the water supply distribution. Nevertheless, consuming reclaimed water has never been a common occurrence. Utilizing treated municipal wastewater for irrigation has been a well-established practice, particularly in regions with arid climates. Incorporating the reuse of wastewater into sustainable water management practices ensures that water can be utilized as an alternative source for various human activities. This can help address the issue of water scarcity and relieve the strain on natural water sources and groundwater (Bixio et al., 2006) (Figure 2.1).

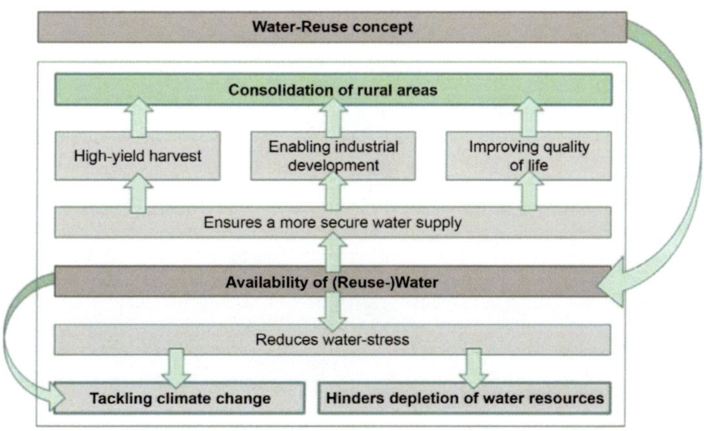

Figure 2.1. *Illustration of water reuse concept.*

Source: W. M. Alley; Creative Commons License.

Various technologies are employed in the treatment of wastewater for reuse. Combining both techniques can ensure that the treated water meets stringent treatment standards and is safe for use, being free from any harmful pathogens. Here are a few common technologies: ultrafiltration, ozonation, aerobic treatment using membrane bioreactors, advanced oxidation, forward osmosis, reverse osmosis, and activated carbon. Certain water-intensive activities can be sustained with lower-quality water. In this scenario, wastewater has the potential to be reused without requiring extensive treatment (Jaramillo & Restrepo, 2017).

2.2.2. Desalinated Water

Desalination is a method used to remove minerals in salty water. In a broader sense, desalination involves the process of eliminating salts and minerals from a specific substance. This is particularly relevant in the context of soil desalination, which poses a significant challenge for the field of agriculture. Desalination processes are used to make salt water, particularly seawater, safe for human intake or irrigation purposes. Brine is produced because of the desalination process. Desalination is a common practice on various seagoing vessels and submarines. There is a significant focus on providing cost-effective clean water for human consumption in modern desalination efforts. In addition to recycled wastewater, rainfall-independent water resources are limited in availability (Yermiyahu et al., 2007).

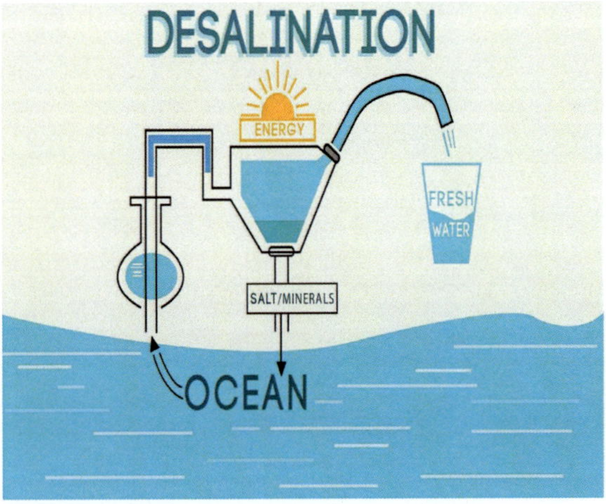

Figure 2.2. *Illustration of water reuse concept.*

Source: Abbas Hassan; Creative Commons License.

Desalinating seawater is typically more expensive than obtaining fresh water coming from surface water or groundwater sources, as well as through water recycling and conservation efforts, primarily due to its higher energy consumption. Unfortunately, such options are not always accessible, and the loss of reserves is a pressing issue on a global scale. There are different methods used in desalination processes, such as thermal techniques like distillation or membrane-based methods like reverse osmosis, which require energy (Reddy & Ghaffour, 2007) (Figure 2.2).

2.3. GROUND WATER RESOURCES

Learning Objectives:

- Understand the factors influencing groundwater resources in desert regions.
- Know the challenges and sociological aspects related to groundwater availability.
- Comprehend the mechanisms and limitations of groundwater replenishment in tropics.
- Explore the significance and management strategies for sustainable groundwater storage.

2.3.1. Groundwater in Desert

Research suggests that the aridity and humidity of a given region are influenced by various factors, including geological climatic belts surrounding the area, and geomorphological events. These factors, particularly when occurred in the recent past, can have a significant impact on the current conditions, such as the prevailing desert regime in North Africa; Geomorphological and geological processes have the potential to significantly impact the climate of certain regions, leading to a lack of synchronization between observed characteristics of the soils, water supply, and more. Furthermore, the prevailing weather conditions (Fan et al., 1997). The lowland regions of northern Kenya encompass a stretch of dry plains that span from Ethiopia and Somalia toward Sudan and the Sahara. These arid lands have remained dry for countless years, dating back to the tertiary era. Lake Rudolph was formed due to a gentle downwarping. There has been a noticeable increase in precipitation and a decrease in aridity in various areas of the region, possibly due to geomorphological uplifts. Geological processes have had a significant impact on the quality and quantity of groundwater.

It is widely acknowledged that during the quaternary epoch, a significant uplift of large areas of the continent resulted in a decrease in the salt level of waters. At the same time, the depressed landmasses experienced an accumulation of salt, leading to the development of groundwater salinity (McKenna & Sala, 2017). Artesian waters hold significant value in regions with dry climates, although their full potential has yet to be realized. In Algeria, shots refer to the outflows of either natural or drilled artesian water, which create lakes and swamps. According to Nace's estimation, a significant amount of water is stored in the Nubian Sandstone and Continental Intercalary Fountain, covering an area of 6.5×10^6 km^3. This storage accounts for approximately 600×10^3 km^3 of water. The current recharge of these reservoirs is negligible, and they were formed during the pluvial period at the end of the Pleistocene, which means that the age of the water is 30,000 to 40,000 years. The sources have been occasionally tapped by hand by the local inhabitants down to depths of almost 150 m, while in the arid parts of South Africa, the water has been traditionally collected by so-called air wells, as it condensed on the surface of limestone blocks (El-Agha et al., 2024).

2.3.2. Groundwater in Wet and Dry Regions

While groundwater resources are generally more abundant in non-arid regions, there can still be challenges with water availability, especially towards the close of the dry season or after a year with inadequate rainfall to replenish the reserves of groundwater. Even wells that are built based on extensive knowledge and expertise can eventually run out of water. In certain regions, even with sufficient groundwater storage, drilled boreholes might be unable to provide water because of the geological structure of the aquifers. On the eastern side of Kariba, for example, nearly half of the approximately 200 boreholes and wells were unable to provide enough water, with some of them completely dry (Yao et al., 2018).

In this and other areas, sociological factors become relevant as the residents who have historically relied on river water are reluctant to consume water from wells, considering it lifeless, some individuals may choose to patiently wait near a sand hole in the dry riverbed for this to fill in both wet and dry regions. In such cases, two distinct groundwater regimes can typically be identified based on the soil profile and the type of aquifer. Certain wells display a heightened sensitivity to changes in precipitation patterns, with groundwater levels in these regions experiencing a rapid increase shortly after the onset of the rainy season. Once the aquifer reaches its maximum saturation, the groundwater level stays relatively stable until the conclusion of the rainy season, sometimes extending a few days beyond. Water resources found in shallow regions, such as riverbeds, sandy aquifers, and intermittent swamps, with impermeable layers near the surface, exhibit a certain behavior (Pahar & Dhar, 2014).

2.3.3. Groundwater Replenishment

The recharge of groundwater in the tropics is severely restricted. There are three potential sources in addition to the water stored over previous geological periods:

- Condensation of atmospheric humidity;
- Rainfall;
- Infiltration of surface water from streams and lakes (Duong & Saphores, 2015).

2.3.3.1. Condensation of Atmospheric Humidity

Condensation of atmospheric Humidity occurs when water vapor in the air transforms into droplets of liquid water as the temperature decreases. The condensed water can seep through the soil and recharge underground water reserves. Nevertheless, the efficacy of this method for groundwater recharge in tropical regions may be constrained by factors like elevated temperatures and humidity levels, potentially leading to a decrease in the frequency and volume of condensation (Castillo et al., 2015).

Practice Problem

Agri Tech, a leading agricultural enterprise, relies on groundwater for irrigation purposes in a region experiencing water stress. To combat declining groundwater levels and ensure sustainable water usage, Agri Tech Company plans to implement a groundwater replenishment project. They aim to allow infiltration of surface water from a nearby river into the groundwater aquifer. The region receives an average annual precipitation of 700 millimeters, and the soil has an infiltration rate of 6 millimeters per hour. Calculate the potential groundwater recharge volume resulting from this infiltration process.

Solutions to Practice Problem

To estimate the potential groundwater recharge volume resulting from the infiltration of surface water from the nearby river into the groundwater aquifer, we first calculate the total annual precipitation volume by multiplying the average annual precipitation by the relevant surface area. Next, we convert the precipitation volume into infiltration volume using the soil infiltration rate. Finally, multiplying the infiltration volume by the number of hours in a day gives us the daily recharge volume. By implementing this groundwater replenishment strategy, Agr Tech aims to enhance groundwater levels, support sustainable agriculture, and mitigate the risk of water shortages in the region.

2.3.3.2. Rainfall

Rainfall plays a crucial role in replenishing groundwater in various regions, regardless of their climate. When rain or snow falls onto the ground, some of it seeps into the soil and eventually makes its way to the water table, replenishing underground water reserves. In regions with tropical climates, where rainfall is plentiful and heavy, this process plays a crucial role in replenishing groundwater storage, especially during the rainy season (Hulme, 1992) (Figure 2.3).

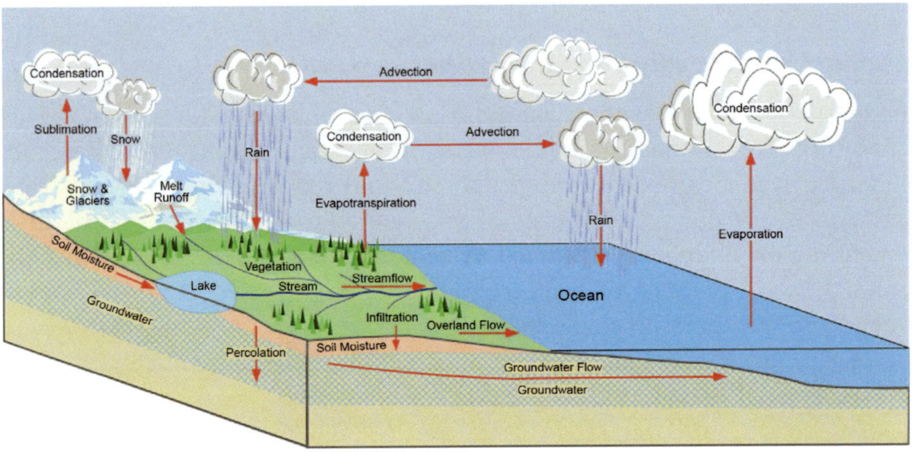

Figure 2.3. *Illustration of the simplified hydrologic cycle.*

Source: O. L. Franke; Creative Commons License.

2.3.3.3. Infiltration of Surface Water From Streams and Lakes

The infiltration of the surface water entering lakes and streams refers to the process by which water from these bodies, like rivers, streams, and lakes, moves into the soil below and eventually reaches groundwater aquifers. This method is essential for restoring groundwater reserves, particularly in areas where sources of surface water are integrated with groundwater systems. In regions with tropical climates, the surface water bodies may change water levels because of seasonal variations in rainfall. Infiltration from such sources plays a crucial role in replenishing groundwater (Hunt et al., 2006).

2.3.4. Ground Water Storage

Groundwater storage pertains to the underground reserves or aquifers that hold water below the Earth's surface. These aquifers are made up of porous rock, sand, gravel, or other materials having the ability to retain and transport water. Groundwater storage plays a critical role in the hydrological cycle and is essential for supplying fresh water to support a wide range of agriculture, human activities, and ecosystems (Döll et al., 2012).

1. **Nature of Aquifers:** There is a wide range of variations in the size, depth, and permeability of aquifers. There are different types of aquifers, some of which are confined among impermeable layers such as rock or clay, whereas others are unrestricted and can directly relate to surface water bodies (Ferguson & Gleeson, 2012).
2. **Recharge and Discharge:** Groundwater storage is replenished through various processes, including rainfall infiltration, surface water infiltration, and artificial recharge methods. Water resources are replenished via natural springs, seepage into rivers and lakes, and extraction by wells (Freeze, 1969).
3. **Storage Capacity:** The capacity of storage for groundwater is influenced by various factors, including the characteristics that define the aquifer material, the rock or sediment's hydraulic conductivity, or the depth of the groundwater table (Cowan, 2001).
4. **Sustainability:** It is crucial to ensure the sustainable management of underground water storage to avoid excessive depletion and extraction of aquifers. Excessive pumping can result in the depletion of water levels, the sinking of land, the infiltration of saltwater in coastal regions, and various other ecological consequences (Placet et al., 2005).
5. **Role in Water Supply:** Groundwater storage is crucial for meeting water demands, especially in areas in which surface water resources are scarce or unpredictable. It plays a crucial role in mitigating the impact of droughts and ensuring a steady flow of water in rivers and streams (Vakilifard et al., 2018).
6. **Monitoring and Management:** It is essential to closely monitor the levels, quality, and recharge rates of groundwater to effectively manage and conserve this valuable resource. Implementing sustainable groundwater management procedures involves the implementation of projects to recharge groundwater, the regulation of pumping rates, and the promotion of water-efficient practices (Cullen et al., 2003).

Groundwater storage is an essential resource that demands careful oversight to guarantee its long-term viability for present and future generations (Figure 2.4).

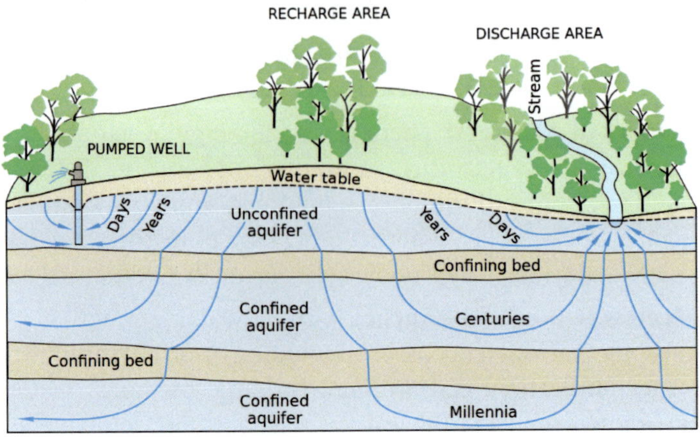

Figure 2.4. *Illustration of the aquifer.*

Source: J. W. Harvey; Creative Commons License.

SUMMARY

- The chapter probes into the diverse array of water resources and their management strategies. It begins by exploring natural sources like surface water bodies and glacier melt, highlighting their vital roles in sustaining ecosystems and human societies.
- Moving on to artificial sources, the chapter discusses innovative methods like wastewater reuse and desalination, crucial for meeting increasing water demands.
- Groundwater resources were another focal point of our discussion, as they represent a significant source of freshwater, particularly in arid regions.
- Moreover, we probed into the mechanisms involved in replenishing groundwater reserves, including the condensation of atmospheric humidity, rainfall, and the infiltration of surface water from streams and lakes.
- Finally, we explored the importance of groundwater storage as a means of buffering against seasonal fluctuations and droughts. By implementing sustainable management practices, we can safeguard this precious resource for future generations.

REVIEW QUESTIONS

1. How do surface water resources benefit human activities?
2. What are the environmental considerations of wastewater reuse?
3. Compare desalination with traditional freshwater sources.
4. What factors influence groundwater replenishment?
5. Evaluate current groundwater management practices.
6. How does groundwater usage differ in arid and wet regions?

MULTIPLE CHOICE QUESTIONS

1. Which of the following is an example of a natural source of freshwater?

a. Wastewater

b. Rainwater

c. Desalinated water

d. Bottled water

2. **Surface water resources primarily include water bodies such as:**
    a. Glaciers
    b. Underground aquifers
    c. Rivers and lakes
    d. Desalination plants
3. **What is one of the artificial sources of usable water mentioned in the chapter?**
    a. Groundwater in deserts
    b. Water from glaciers
    c. Wastewater reuse
    d. Groundwater storage
4. **Groundwater resources can be found:**
    a. Only in desert regions
    b. Exclusively in wet regions
    c. In both desert and wet regions
    d. None of the above
5. **Which process is involved in replenishing groundwater according to the chapter?**
    a. Desalination
    b. Bottled water production
    c. Rainfall
    d. Wastewater reuse

Answers to Multiple Questions

1. (b); 2. (c); 3. (c); 4. (c); 5. (c)

REFERENCES

1. Adewumi, J. R., Ilemobade, A. A., & Van Zyl, J. E. (2010). Treated wastewater reuse in South Africa: Overview, potential and challenges. *Resources, Conservation and Recycling, 55*(2), 221–231. https://doi.org/10.1016/j.resconrec.2010.09.007.
2. Bixio, D., Thoeye, C., De Koning, J., Joksimovic, D., Savic, D., Wintgens, T., & Melin, T. (2006). Wastewater reuse in Europe. *Desalination, 187*(1–3), 89–101. https://doi.org/10.1016/j.desal.2005.04.077.
3. Castillo, J. E., Weibel, J. A., & Garimella, S. V. (2015). The effect of

relative humidity on dropwise condensation dynamics. *International Journal of Heat and Mass Transfer, 80*, 759–766. https://doi.org/10.1016/j.ijheatmasstransfer.2014.09.089.

4. Cowan, N. (2001). Metatheory of storage capacity limits. *Behavioral and Brain Sciences, 24*(1), 154–176. https://doi.org/10.1017/S0140525X01003922.
5. Cullen, J., Joyce, J., Hassall, T., & Broadbent, M. (2003). Quality in higher education: From monitoring to management. *Quality Assurance in Education, 11*(1), 5–14. https://doi.org/10.1108/09684880310461715.
6. Döll, P., Hoffmann-Dobrev, H., Portmann, F. T., Siebert, S., Eicker, A., Rodell, M., & Scanlon, B. R. (2012). Impact of water withdrawals from groundwater and surface water on continental water storage variations. *Journal of Geodynamics, 59*, 143–156. https://doi.org/10.1016/j.jog.2012.03.010.
7. Duong, K., & Saphores, J. D. M. (2015). Obstacles to wastewater reuse: An overview. *Wiley Interdisciplinary Reviews: Water, 2*(3), 199–214. https://doi.org/10.1002/wat2.1070.
8. El-Agha, D. E., Molle, F., Metwally, M. I., Emara, S. R., Shalby, A., Armanuos, A. M., & Gado, T. A. (2024). Toward sustainable management of groundwater in the deserts of Egypt. *Hydrogeology Journal, 32*(3), 663–678. https://doi.org/10.1007/s10040-023-02430-1.
9. Epstein, S., & Mayeda, T. (1953). Variation of O^{18} content of waters from natural sources. *Geochimica et Cosmochimica Acta, 4*(5), 213–224. https://doi.org/10.1016/0016-7037(53)90051-9.
10. Fan, Y., Duffy, C. J., & Oliver Jr, D. S. (1997). Density-driven groundwater flow in closed desert basins: Field investigations and numerical experiments. *Journal of Hydrology, 196*(1–4), 139–184. https://doi.org/10.1016/S0022-1694(96)03367-1.
11. Ferguson, G., & Gleeson, T. (2012). Vulnerability of coastal aquifers to groundwater use and climate change. *Nature Climate Change, 2*(5), 342–345. https://doi.org/10.1038/nclimate1413.
12. Freeze, R. A. (1969). The mechanism of natural ground-water recharge and discharge: 1. One-dimensional, vertical, unsteady, unsaturated flow above a recharging or discharging ground-water flow system. *Water Resources Research, 5*(1), 153–171. https://doi.org/10.1029/WR005i001p00153.
13. Hulme, M. (1992). Rainfall changes in Africa: 1931–1960 to 1961–1990. *International Journal of Climatology, 12*(7), 685–699. https://doi.org/10.1002/joc.3370120704.

14. Hunt, R. J., Strand, M., & Walker, J. F. (2006). Measuring groundwater–surface water interaction and its effect on wetland stream benthic productivity, Trout Lake watershed, northern Wisconsin, USA. *Journal of Hydrology, 320*(3–4), 370–384. https://doi.org/10.1016/j.jhydrol.2005.07.006.
15. Jaramillo, M. F., & Restrepo, I. (2017). Wastewater reuse in agriculture: A review about its limitations and benefits. *Sustainability, 9*(10), 1734. https://doi.org/10.3390/su9101734.
16. McKenna, O. P., & Sala, O. E. (2017). Groundwater recharge in desert playas: Current rates and future effects of climate change. *Environmental Research Letters, 13*(1), 014025. https://doi.org/10.1088/1748-9326/aa9b60.
17. McKinney, D. C., & Cai, X. (2002). Linking GIS and water resources management models: An object-oriented method. *Environmental Modelling & Software, 17*(5), 413–425. https://doi.org/10.1016/S1364-8152(02)00010-3.
18. Nagrodski, A., Raby, G. D., Hasler, C. T., Taylor, M. K., & Cooke, S. J. (2012). Fish stranding in freshwater systems: Sources, consequences, and mitigation. *Journal of Environmental Management, 103*, 133–141. https://doi.org/10.1016/j.jenvman.2012.03.007.
19. Naidoo, S., & Olaniran, A. O. (2014). Treated wastewater effluent as a source of microbial pollution of surface water resources. *International Journal of Environmental Research and Public Health, 11*(1), 249–270. https://doi.org/10.3390/ijerph110100249.
20. Pahar, G., & Dhar, A. (2014). A dry zone-wet zone-based modeling of surface water and groundwater interaction for the generalized ground profile. *Journal of Hydrology, 519*, 2215–2223. https://doi.org/10.1016/j.jhydrol.2014.10.016.
21. Park, Y. S., Céréghino, R., Compin, A., & Lek, S. (2003). Applications of artificial neural networks for patterning and predicting aquatic insect species richness in running waters. *Ecological Modelling, 160*(3), 265–280. https://doi.org/10.1016/S0304-3800(02)00257-4.
22. Placet, M., Anderson, R., & Fowler, K. M. (2005). Strategies for sustainability. *Research-Technology Management, 48*(5), 32–41. https://doi.org/10.1080/08956308.2005.11657323.
23. Reddy, K. V., & Ghaffour, N. (2007). Overview of the cost of desalinated water and costing methodologies. *Desalination, 205*(1–3), 340–353. https://doi.org/10.1016/j.desal.2006.03.558.

24. Shreve, R. L. (1972). Movement of water in glaciers. *Journal of Glaciology, 11*(62), 205–214. https://doi.org/10.3189/S0022143000022736.
25. Snelder, T. H., & Biggs, B. J. (2002). Multiscale river environment classification for water resources management 1. *JAWRA Journal of the American Water Resources Association, 38*(5), 1225–1239. https://doi.org/10.1111/j.1752-1688.2002.tb04344.x.
26. Vakilifard, N., Anda, M., Bahri, P. A., & Ho, G. (2018). The role of water-energy nexus in optimizing water supply systems–Review of techniques and approaches. *Renewable and Sustainable Energy Reviews, 82*, 1424–1432. https://doi.org/10.1016/j.rser.2017.05.125.
27. Weatherhead, E. K., & Howden, N. J. K. (2009). The relationship between land use and surface water resources in the UK. *Land Use Policy, 26*, S243–S250. https://doi.org/10.1016/j.landusepol.2009.09.013.
28. Yao, Y., Tian, Y., Andrews, C., Li, X., Zheng, Y., & Zheng, C. (2018). Role of groundwater in the dryland ecohydrological system: A case study of the Heihe River Basin. *Journal of Geophysical Research: Atmospheres, 123*(13), 6760–6776. https://doi.org/10.1029/2018JD028432.
29. Yermiyahu, U., Tal, A., Ben-Gal, A., Bar-Tal, A., Tarchitzky, J., & Lahav, O. (2007). Rethinking desalinated water quality and agriculture. *Science, 318*(5852), 920–921. https://doi.org/10.1126/science.1146339.

CHAPTER 3

HYDROLOGY PRINCIPALS AND PROCESSES

LEARNING OBJECTIVES

At the end of this chapter, readers will be able to:

1. Understand the hydrologic cycle and its significance in water movement.
2. Recognize key components of hydrologic systems and their interactions.
3. Understand atmospheric and ocean circulation patterns and their impact on hydrology.
4. Know the formation mechanisms and types of precipitation.
5. Comprehend the factors influencing evaporation and methods for estimating it.
6. Understand percolation and groundwater recharge processes.

INTRODUCTORY EXAMPLE

HydroFlow Dynamics, a pioneering hydrology research firm, specializes in understanding and harnessing the power of water systems for sustainable development. With expertise in hydrological principles and processes, HydroFlow Dynamics collaborates with governments and industries to address complex water-related challenges.

In a recent project, HydroFlow Dynamics was engaged by a coastal city facing increasing risks of flooding due to rising sea levels and changing precipitation patterns. Leveraging advanced hydrological models, the team at HydroFlow Dynamics simulated various scenarios to assess flood vulnerability and identify adaptation measures.

Through meticulous analysis of ocean and atmospheric circulation patterns, coupled with detailed precipitation forecasting, HydroFlow Dynamics provided actionable insights to enhance the city's resilience to climate-induced flooding. Recommendations included the implementation of green infrastructure, such as rain gardens and permeable pavements, and the adoption of smart drainage systems to mitigate flood risks while preserving natural habitats.

HydroFlow Dynamics continues to lead the way in innovative hydrology research, empowering communities and industries to thrive in an ever-changing water landscape.

UNIT INTRODUCTION

Water, a vital element for all life forms, has a profound impact on shaping the Earth's terrain, maintaining ecosystems, and ensuring the survival of human societies. Water is constantly in motion, transitioning, and changing through an intricate web of processes, from the expansive oceans to the tiniest droplets in the air. Hydrology, an interesting area of study, offers valuable insights into the understanding and management of Earth's water resources (Salas et al., 2014).

Within this chapter, the core principles and mechanisms of hydrology, which is the scientific study of water's movement, distribution, and quality on the planet will be explored. By delving into the fundamentals of hydrology, such as the hydrologic cycle and different hydrological systems, one can gain a deeper understanding of the complex mechanisms that regulate the Earth's water resources (Edwards et al., 2015).

This unit offers a thorough examination of the various factors that impact the distribution and movement of water, ranging from atmospheric and ocean circulation to the formation and types of precipitation. In addition, the processes of evaporation and methods for estimating it, along with the significant role of percolation and groundwater in the hydrological cycle will be explored (Shaw et al., 2019).

3.1. ESSENTIALS OF HYDROLOGY

Learning Objectives:

- Understand the scope and significance of hydrology as a scientific discipline.
- Explore the components and processes of the hydrologic cycle.
- Identify the key features and dynamics of atmospheric and ocean circulation.

3.1.1. Definition of Hydrology

Hydrology is the scientific study of the Earth's water, encompassing its presence, movement, and distribution, as well as its physical and chemical characteristics, and its interaction with the environment, including its connection to living organisms. Hydrology includes the complete life cycle of water on Earth (Wagener et al., 2010).

3.1.2. The Hydrologic Cycle

The hydrologic cycle, which consists of several continuous processes as shown in Figure 3.1, is the main subject of hydrology. Water vapor, which is carried throughout the planet by atmospheric circulation, is created when water evaporates from the oceans and land surfaces. When water vapor in the atmosphere condenses and falls to the Earth's surface, precipitation happens. Various processes can occur when water is present, such as vegetation intercepting it, flowing over the ground surface, seeping into the ground, moving through the soil as subsurface flow, and eventually being discharged as surface runoff (Loaiciga et al., 1996).

Evaporation from the land surface is a process that includes the direct evaporation from soil and vegetation surfaces, along with the transpiration through plant leaves. These processes are commonly referred to as evapotranspiration. Water that has infiltrated may seep down further to replenish underground water sources (Horton, 1933).

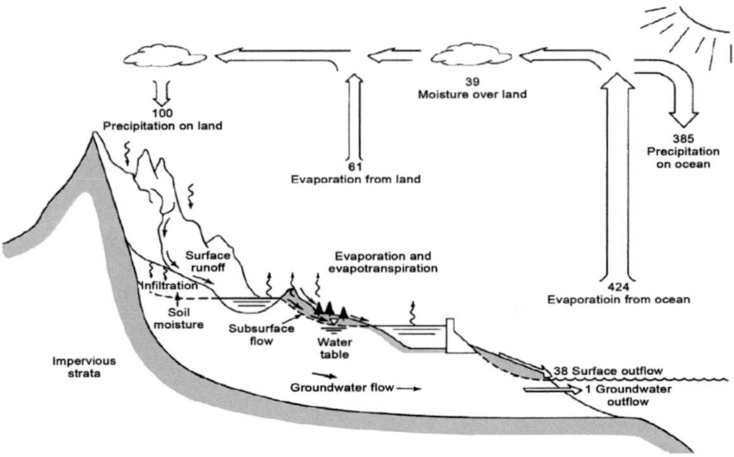

Figure 3.1. *Illustration of the hydrologic cycle, where the world annual average water balance is expressed in units in relation to the land surface precipitation rate of 100.*

Source: Chow; Creative Commons License.

During this runoff process, water can be stored in various natural reservoirs, including glaciers, ice sheets, snow, streams, soils, lakes, sediments, rock, and vegetation formations. The evaporation from these reservoirs disrupts the global hydrologic cycle, creating smaller cycles of varying scales. The runoff is widely recognized as a key component of the global hydrologic cycle, yet its accuracy is often surrounded by considerable uncertainty (Linton, 2008).

3.1.3. Hydrologic Systems

In their research, Chow and Maidment defined a hydrologic system. They defined it as a boundary-containing structure or volume in space that can take in water and other inputs. The system internally processes these inputs and generates these outputs. The organization or distribution in space is the comprehensive network of flow paths by which the water can move from its starting point to its endpoint in the system (Manabe, 1969).

A surface that encloses a three-dimensional specified volume or structure is called a boundary. A working medium is fed into the system as input. It then interacts with the other media and the structure to become an output. The system is influenced by various processes involving water, air, and heat energy, which play a significant role in hydrologic analysis (O'Connell & Todini, 1996).

Figure 3.2 shows the three subsystems that make up the global hydrologic cycle: the surface water system, the subsurface water system, and the atmosphere. An additional illustration relates to the storm-rainfall-runoff cycle within a watershed, which can be represented as a hydrologic system. While stream flow at the watershed's exit is the intended result, the data presented shows patterns of rainfall throughout the watershed. The watershed divide establishes the boundary, which reaches horizontal planes in both an upward and downward direction (Manabe et al., 1975).

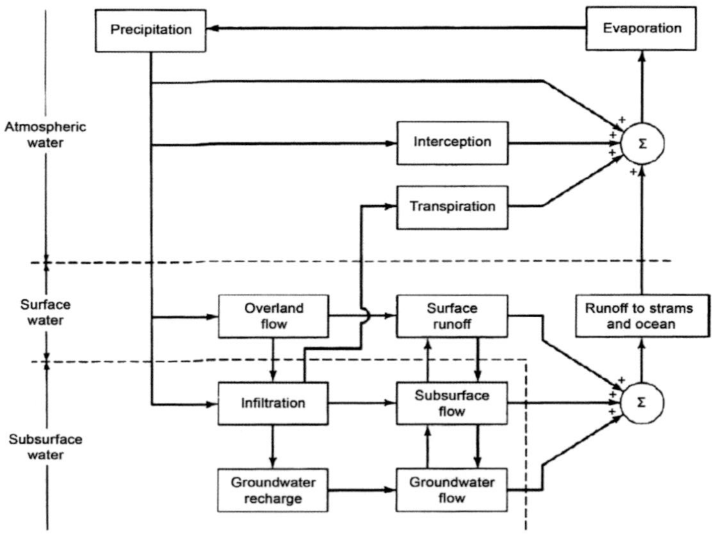

Figure 3.2. *Illustration of the global hydrologic system using block diagrams.*

Source: Philip, B.; Creative Commons License.

The terms watersheds, drainage basins, and catchments are all used interchangeably to describe the topographic area that gathers and discharges surface stream flow through one outlet or mouth. Catchments are commonly known as small drainage basins, although there are no specific area limits that have been set (Leray et al., 2016).

3.1.4. Ocean and Atmospheric Circulation

Earth's atmospheric circulation is a complex process that is impacted by numerous variables. Several factors contribute to the phenomenon, including variations in temperature at different altitudes, the rotation of the Earth, and differences in heat and pressure between land and water. Several variables, including variations in solar heating, the inclination of the earth's axis, the distribution of water and land, the mechanisms of fluid flow in the atmosphere, and the Coriolis Effect, affect the movement of the atmosphere (Kousky et al., 1984).

The earth's rotation and patterns of global pressure are linked to the movement of the atmosphere, which is driven by thermal causes. The atmospheric circulation on a nonrotating sphere would be represented by Figure 3.3. The movement of air near the equator involves upward motion, followed by a journey in the upper atmosphere towards the poles. After cooling down, it descends into the lower atmosphere and makes its way back towards the equator. This phenomenon is known as the Hadley circulation.

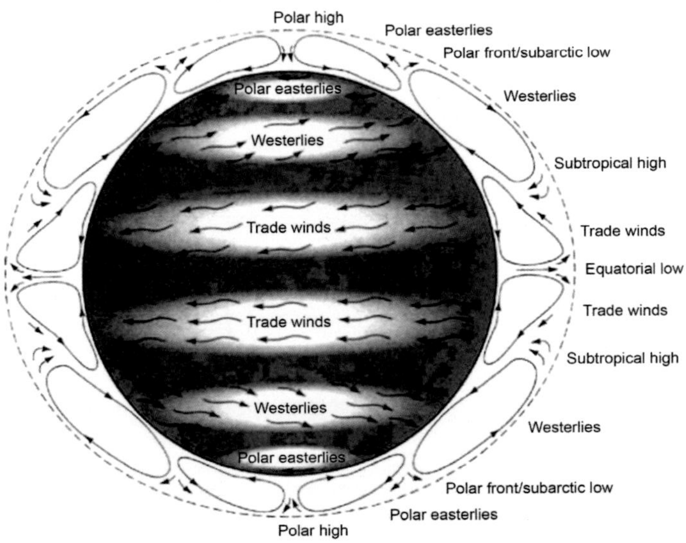

Figure 3.3. *Description of the simplified circulation of the atmosphere near the Earth's surface, highlighting the main pressure areas and wind belts.*

Source: Daniel, P.; Creative Commons License.

The earth's rotation from west to east modifies the pattern of circulation. The air's radius rapidly decreases as it travels closer to the poles along the earth's axis of circulation. To ensure the preservation of angular momentum, the velocity of air experiences an increase to the land surface, resulting in the creation of a westerly airflow. On the other hand, when a ring of air moves towards the equator, it creates an easterly airflow. The Coriolis force is responsible for causing these alterations in wind direction and speed (Winton, 2003).

The pattern of atmospheric circulation is characterized by three cells in each hemisphere. The tropical cell operates by causing heated air to rise at the equator, then move towards the poles at higher altitudes, where it cools down and eventually descends towards the ground around latitude 30°. Close to the surface, it splits into two branches, with one branch heading towards the equator and the other towards the pole. The air in the polar cell rises at 60° latitude and travels at greater elevations towards the poles before cooling and falling to 60° close to the surface of the earth. Friction between the center cell and the other two cells pushes its surface toward the pole. In the mid-latitudes, this movement leads to the establishment of the dominant westerly airflow (Stocker & Wright, 1996). The arrangement of ocean and landmasses across the Earth's surface, combined with their different thermal characteristics, results in spatial disparities in atmospheric circulation. Annually, the thermal equator's movement is affected by the Earth's orbit around the sun, leading to a consistent variation in the three-cell circulation pattern (Bryan et al., 1975).

By increasing the oscillation, there is more chance of increased air exchanges between neighboring cells. This could potentially lead to a greater number of years of flooding. Furthermore, monsoons have the potential to extend further into countries like India and Australia. Given a reduced oscillation, it is conceivable that a significant buildup of high pressure could occur around 30° latitude, leading to prolonged periods of dry weather. Due to the complexity of atmospheric circulation, it is only possible to identify the general pattern.

There are different zones separated vertically inside the atmosphere. The troposphere, which varies in height from roughly 8 km at the poles to 16 km at the equator, is where the atmospheric circulation mentioned above takes place. The troposphere's temperature drops with altitude at a pace that depends on the atmosphere's moisture content (Jansen, 2017).

For dry air, there is a specific rate of decrease known as the dry adiabatic lapse rate, which is roughly 9.8°C/km. When the air rises and cools, some of the vapor in it condenses, releasing heat into the surrounding air. This causes the saturated adiabatic lapse rate to be lower, approximately 6.5°C/km. These figures for lapse rates can vary significantly with altitude (Smith et al., 2006).

The boundary between the troposphere and the stratosphere is known as the tropopause. Near the tropopause, there are significant shifts in temperature and pressure that give rise to powerful and narrow air currents called jet streams, which can reach speeds of 15 to 50 m/s (30 to 100 m/h). They span across vast distances and play a significant role in the movement of air masses.

3.2. PRECIPITATION (RAINFALL)

Learning Objectives:

- Understand the various forms and processes of precipitation in hydrology.
- Know the mechanisms and factors influencing rainfall variability over different regions.
- Explore the dynamics of disposal of rainfall on a watershed and its implications.

Precipitation refers to any form of water, whether it is in a liquid or frozen state that condenses in the atmosphere and then falls back to the earth. There are various forms of precipitation, including snow, sleet, and rain. One of the three main components of the global water cycle is precipitation, which occurs alongside evaporation and condensation (Lee & Heo, 2011). Precipitation, such as rain, sleet, hail, and snow, typically happens due to adiabatic cooling. Adiabatic cooling refers to the process in which an air mass ascends, expands, and becomes cooler. Because cool air has a lower capacity to store water compared to warm air, when an air mass is cooled, it can become supersaturated with water vapor. This results in the development of clouds and condensed water droplets, which then precipitate on the ground (Sun et al., 2018).

Various forms of air mass movement can result in precipitation with distinct attributes:

1. **Orographic:** When moist air meets mountains, it is pushed upwards, causing it to cool and resulting in rainfall on the side of the mountains facing the wind. In areas sheltered by mountains, rainfall is usually minimal due to the warm, dry air descending from higher altitudes (Trenberth et al., 2003).
2. **Frontal:** It is common for precipitation to develop where air masses with varying temperatures and densities meet. When a cold front moves under warmer air, it results in intense rainfall over a short period. On the other hand, when a warm front displaces cooler air, it leads to longer periods of rainfall with lower intensity. Tropical cyclones also referred to as hurricanes, can unleash significant volumes of rainfall within a brief period (Javanmard et al., 2010).
3. **Convective:** Warm afternoons may cause the air close to the surface to heat up and travel higher, producing brief but heavy downpours that usually take the form of thunderstorms.

Radar or rain gauges can be used to estimate the amount of precipitation. A hyetograph is a graph showing the amount of precipitation over time.

Precipitation is stated as total depth over an event (e.g., mm) or as depth per unit of time (e.g., mm/day). When a droplet of waterfalls, it has two options: it can either land directly on the ground or contact a surface that is covered in vegetation, such as a leaf on a tree. Moisture trapped on vegetation can take two routes: it can either trickle down the plant or reach the ground (known as stem flow), or it is more prone to evaporate from the plant without ever making it to the ground. After it falls to the ground, the water droplet has two options: it can either flow across the surface or penetrate the soil and add to the soil moisture (Odening et al., 2007).

Porosity is a crucial factor in determining the soil's capacity to retain water. The calculation of porosity involves determining the ratio between the volume of pores and the total volume. When the pore space is filled with water instead of air, the soil is saturated. In an introduction to water resource science, it is often mentioned that the soil near the surface is typically not saturated. The volumetric moisture content, which is the volume of water divided by the total volume, can range from zero to porosity (Nesbitt et al., 2006).

3.2.1. Precipitation Types and Formation

While precipitation encompasses rainfall, snowfall, hail, and sleet, the focus is primarily on rainfall. Figure 3.4 provides a visual representation of the process of water droplet formation in clouds. Condensation occurs in the atmosphere when tiny particles, known as condensation nuclei, are present. These particles, which can range in size from (10^{-3} – 10μm), are typically made up of salt or dust (Rosinski, 1974).

These particles are referred to as aerosols. During the early stages of condensation, tiny droplets or ice particles are suspended in the air due to the movement of air molecules. After droplets are created, they also serve as condensation nuclei. These minuscule water droplets naturally tend to repel one another. However, when an electric field is present in the atmosphere, they start to attract one another. Eventually, these droplets become heavy enough, around 0.1 mm in size, to fall through the atmosphere. Various droplets undergo different processes, including evaporation, size reduction, and size increase through impact and aggregation (Bourgouin, 2000).

Understanding the process of precipitation formation involves the lifting of an air mass in the atmosphere, which subsequently cools and results in the condensation of moisture. Three primary mechanisms contribute to the upward movement of air masses: orographic lifting, frontal lifting, and convective lifting. Figure 3.5 shows the process of frontal lifting, which happens when warm air is lifted over cooler air due to the passage of a front.

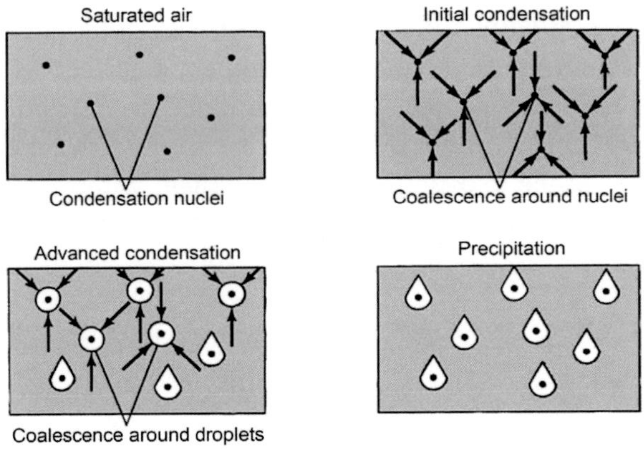

Figure 3.4. *Illustration of precipitation formation.*

Source: Shimon, C.; Creative Commons License.

On the other hand, Figure 3.6 shows the phenomenon of orographic lifting, where an air mass ascends over a mountain range. Lastly, Figure 3.7 illustrates the process of convective lifting, where air is pulled upwards due to convective activity, like that of a thunderstorm cell (Volford et al., 2007).

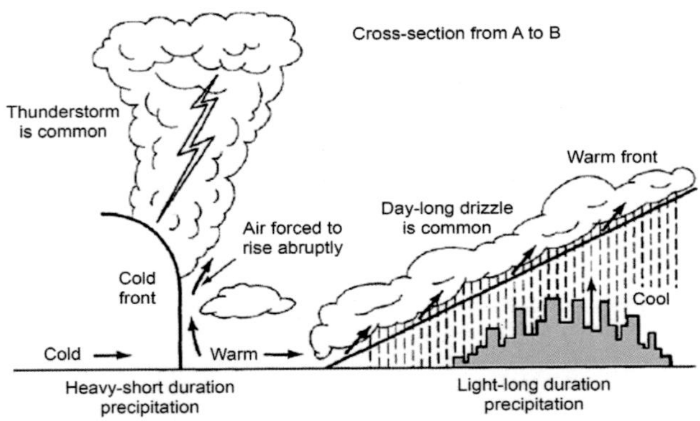

Figure 3.5. *Illustration of cyclonic storms in mid-latitude.*

Source: P. Chow; Creative Commons License.

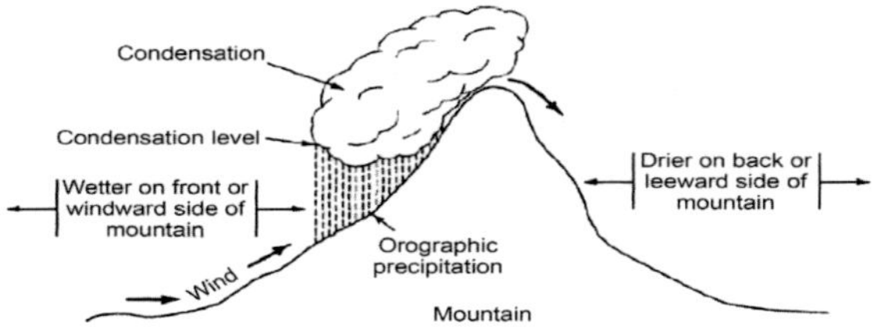

Figure 3.6. *Illustration of orographic storm.*

Source: Hustler, J.; Creative Commons License.

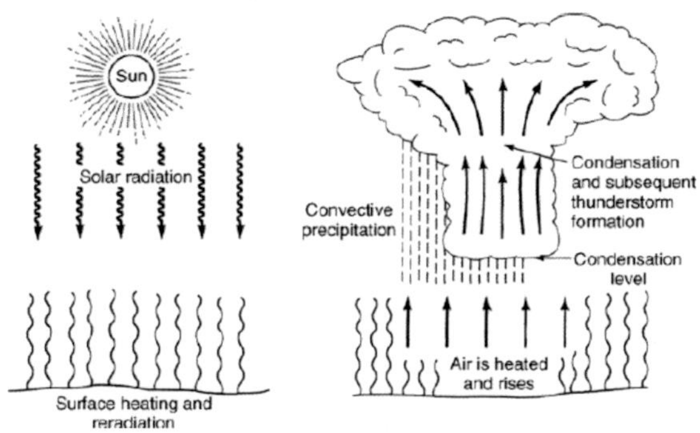

Figure 3.7. *Illustration of convective storm.*

Source: David, B.; Creative Commons License.

3.2.2. Rainfall Variability

Maps of rainfall variability show the annual variations in precipitation. For example, a location will often get more constant rainfall from year to year if its rainfall variability is smaller. This kind of rainfall pattern is found in many coastal places (Shahid, 2010). Conversely, increased rainfall variability indicates a greater possibility of erratic rainfall, with torrential downpours in some years and little to none in others. Precipitation is one of the main inputs used to calculate the runoff from a watershed and the resulting stream flow. Because of local causes and the overall pattern of air circulation, rainfall varies in space and time (Le Barbé et al., 2002).

3.2.3. Disposal of Rainfall on a Watershed

The region of land that empties into a stream at a specific point is known as a watershed. A watershed is the site of the different surface water processes that make up the hydrologic cycle. A schematic representation of how rainfall gets disposed of on a watershed during a storm is shown in Figure 3.8. The pace at which water flows through each operation or is added to storage, as a function of time, is shown in this figure (Frumin et al., 2017). During the onset of a storm, a significant amount of rainfall is responsible for surface storage. As the water seeps into the ground, the storage of soil moisture begins. It is normal practice to use both detention and retention storage. Retention storage is typically held for an extended duration and gradually decreases due to evaporation, while detention storage is designed for shorter periods and is reduced by the flow of water from the storage area (Jones et al., 1984).

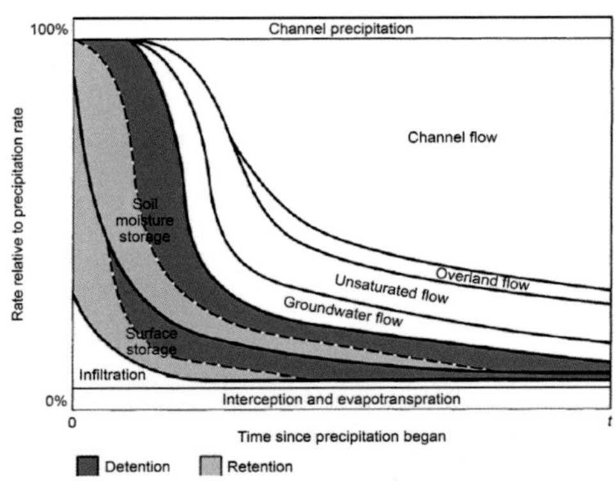

Figure 3.8. *Schematic illustration of the disposal of precipitation during a storm on a watershed.*

Source: Chow; Creative Commons License.

3.3. EVAPORATION

Learning Objectives:

- Understand the process of evaporation and its significance in hydrology.
- Understand the energy balance and aerodynamic method for calculating evaporation.

The transformation of water from its liquid form to a gaseous state is called evaporation. This process can happen from various sources, such as water bodies, saturated soils, or unsaturated surfaces. Figure 3.9 provides a visual representation of the evaporation process. Various events are occurring above the water's surface. Initially, water molecules such as water vapor are consistently present above liquid water (Penman, 1956).

Furthermore, there are a few other molecules:

- Two oxygen atoms bonded together, creating an oxygen molecule; and
- Two nitrogen atoms joined together, resulting in a nitrogen molecule.

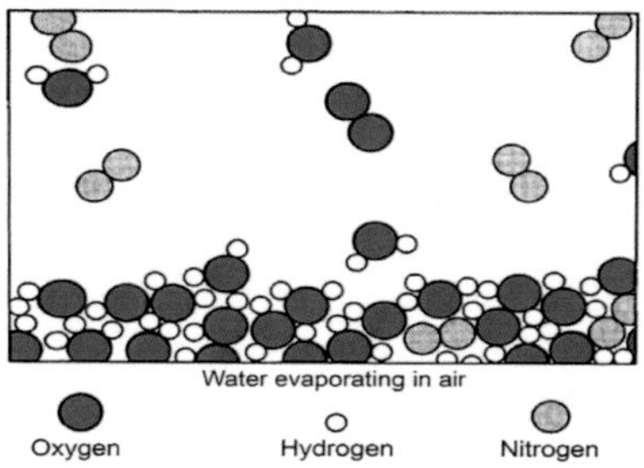

Figure 3.9. *Illustration of evaporation.*

Source: Feynman; Creative Commons License.

The gas above the water surface known as the air is primarily composed of nitrogen and oxygen, with traces of additional elements such as argon, carbon dioxide, and water vapor. Water molecules are in a state of constant motion. One on the surface is occasionally knocked away. In other terms, the water evaporates or vanishes molecule by molecule.

Calculating evaporation in hydrologic analysis and design plays a crucial role in water supply design, especially in the design and operation of reservoirs. The energy source needed to produce the latent heat of vaporization and the ability to move water vapor away from the evaporative surface are two important variables that affect the evaporation process (Burt et al., 2005).

Latent heat refers to the heat that is released or absorbed during a phase transition, such as from a solid to a liquid or from a liquid to a gas. The latent heat of vaporization (l_v) is the amount of heat released when liquid water turns into water vapor (Chen et al., 2013). It can be calculated using the formula l as $l_v = 2{:}501 \times 10^6 — 2370T$, where 'T' represents the temperature in degree Celsius (°C) and lv is measured in joules (J) per kilogram. Evaporation is commonly calculated using two methods: the aerodynamic method and the energy balance method. The energy balance method, which considers the thermal energy balance of a hydrologic system, is one strategy to take into consideration. The aerodynamic method is an alternative strategy that concentrates on the system's ability to transfer away from an open surface. A graphic depiction of the radiation and heat regulation in the atmosphere and at the surface of the earth is shown in Figure 3.10, which also lists the various components and their corresponding values (Klots, 1991).

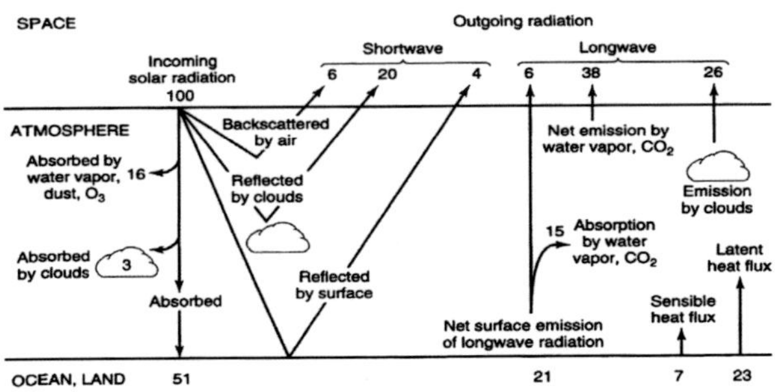

Figure 3.10. *Illustration of heat and radiation balance at the surface of the planet and in the atmosphere.*

Source: Arthur Lions; Creative Commons License.

3.3.1. Energy Balance Method

Examine the evaporation pan with the defined control volume displayed in Figure 3.11. Water has densities of ρ_w in the liquid phase and ρ_a in the vapor phase in this control volume, respectively. It is necessary to write the continuity equation for each phase. The mass flow rate of evaporation is represented by $dB/dt = \dot{m}_v$, the extensive property β = m (mass of liquid water), and the intense property β = 1. Next, the liquid phase's continuity is:

$$-\dot{m}_v = \frac{d}{dt}\int_{CV} \rho_w d\forall + \sum_{CS} \rho_w \mathbf{V}\cdot\mathbf{A} \tag{1}$$

Due to the pan's impermeable sides, there is no movement of liquid water across the control surface, resulting in a constant storage rate (Mehdipour et al., 2010).

$$\frac{d}{dt}\int_{CV} \rho_w d\forall = \rho_w A \frac{dh}{dt} \tag{2}$$

where; 'h' is the water depth; and 'A' is the pan's cross-sectional area. Eqn. (2) can be substituted into Eqn. (1) to obtain:

$$-\dot{m}_v = \rho_w A \frac{dh}{dt} \tag{3}$$

$$\dot{m}_v = \rho_w AE \tag{4}$$

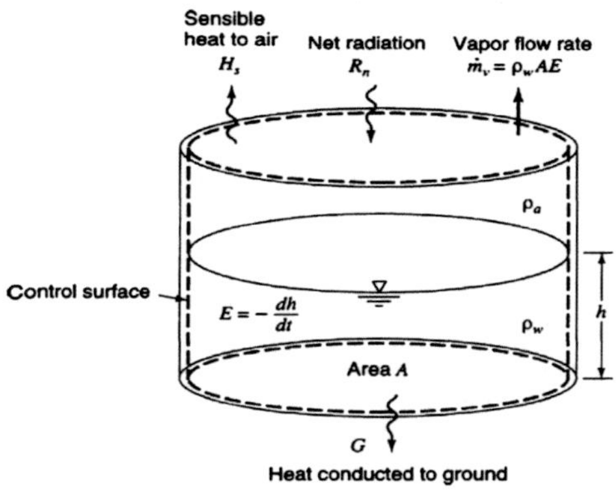

Figure 3.11. *Control volume defined for continuity and energy equation development for an. evaporation pan*

Source: Denis Stone, Creative Commons License.

where; the evaporation rate $E = -dh/dt$

In the vapor phase, the mass of water vapor is the extensive property $= m_v(mass\ of\ water\ vapor)$, the intensive property is $\beta = q_v$, and the $db/dt = \dot{m}_v$. Sustaining the vapor phase:

$$\dot{m}_v = \frac{d}{dt}\int_{CV} q_v \rho_a d\forall + \sum_{CS} q_v \rho_a \mathbf{V}\cdot\mathbf{A} \tag{5}$$

When there is a constant flow over the pan, the water vapor's time derivative in the control volume equals:

$$\rho_w AE \sum_{CS} q_v \rho_a \mathbf{V}\cdot\mathbf{A} \tag{6}$$

This is the pan's continuity equation. It is possible to rearrange Eqn. (6) to produce:

$$E = \left(\frac{1}{\rho_w A}\right) \sum_{CS} q_v \rho_a \mathbf{V}\cdot\mathbf{A} \tag{7}$$

Next, let's look at the heat energy equation:

$$\frac{dH}{dt}-\frac{dW_s}{dt}=\frac{d}{dt}\int_{CV}\left(e_u+\frac{1}{2}V^2+gz\right)\rho_w d\forall+\sum_{CS}\left(\frac{p}{\rho_w}+e_u+\frac{1}{2}V^2+gz\right)\rho_w \mathbf{V}\cdot\mathbf{A} \quad (8)$$

When considering the density of water as ρ_w, the rate of heat input from an external source as dH/dt, and the absence of work done by the system as $dW_s/dt = 0$ ¼, one can simplify the equation. Additionally, considering the specific internal heat energy of the water as , the water in the pan having V = 0, and the small rate of change of the water surface elevation (dz/dt), the equation is reduced to:

$$\frac{dH}{dt}=\frac{d}{dt}\int_{CV} e_u\rho_w d\forall \quad (9)$$

One way to quantify the amount of heat coming from external sources is by expressing it as:

$$\frac{dH}{dt}=R_n-H_s-G \quad (10)$$

where; 'G' is the ground heat flow to the ground surface; 'H_s' is the sensible heat to the air stream given by the water; and 'R_n' is the net radiation flux (watts per meter squared). The amount of radiation entering the surface at any given time is known as net radiation flux (Tanner & Pelton, 1960).

The primary energy source for water evaporation at the surface of the planet is the net radiation flux, which is defined as the difference between the radiation emitted and the radiation absorbed, or $R_i(1-\alpha)$, where R_i is the incident radiation and a is the fraction of radiation reflected, called the albedo), and that emitted. R_e:

$$R_n = R_i(1-\alpha)-R_e \quad (11)$$

The Stefan-Boltzmann law determines the amount of radiation released.

$$R_e = e\sigma T_p^4 \quad (12)$$

The surface emissivity is represented by e, and the surface absolute temperature in degrees Kelvin is represented by T_p, and the Stefan-Boltzmann constant is given by $s = (5.67 \times \frac{10^{-8}W}{m^2}.K^4)$.

The only way the heat stored in the control volume can vary, given a fixed temperature of the water in the control volume, is by altering the internal energy of the water vaporized $l_v\dot{m}_v$ where l_v represents the latent heat of vaporization (Irmak et al., 2014).

$$\frac{d}{dt}\int_{CV} e_u \rho_w d\forall = l_v \dot{m}_v \tag{13}$$

Eqns. (10) and (13) can be substituted into Eqn. (9) to obtain:

$$R_n - H_s - G = l_v \dot{m}_v \tag{14}$$

The equation for evaporation's energy balance is obtained by solving for E, which stands for energy balance:

$$E = \frac{1}{l_v \rho_w}(R_n - H_s - G) \tag{15}$$

The rate at which all incoming net radiation is absorbed by evaporation, or the evaporation rate E_r can be computed by assuming that the sensible heat flow H_s and the ground heat flux G are both zero (Spittlehouse & Black, 1980).

$$E_r = \frac{R_n}{l_v \rho_w} \tag{16}$$

Practice Problem

Visualize you are part of a team at AquaTech, a renowned water technology company known for its innovative solutions in hydrology and water resource management. Your team is tasked with determining the rate of evaporation from a water surface using the energy balance method. Given data on net radiation flux, sensible heat flux, ground heat flux, and the latent heat of vaporization, your objective is to calculate the rate of evaporation. This challenge is crucial for AquaTech's ongoing projects focused on optimizing water usage and understanding evaporation rates in various environments.

Solutions to Practice Problem

To address this challenge, your team at AquaTech applies its expertise in hydrology principles and processes, utilizing the energy balance equation for evaporation. By analyzing the provided data and employing AquaTech's advanced methodologies, you accurately calculate the rate of evaporation from the water surface. This solution showcases AquaTech's commitment to leveraging cutting-edge technology and scientific knowledge to tackle complex water-related issues, reinforcing its position as a leader in the field of water resource management.

3.3.2. Aerodynamic Method

The aerodynamic approach considers the ability to transfer water vapor away from the water's surface, as was previously mentioned. The humidity gradient in the air near the surface and the wind speed across the surface has an impact on this. By integrating the formula for mass and momentum transmission in the air, these processes can be investigated. When analyzing the control volume shown in Figure 3.12, it is important to consider the vapor flux $\dot{m}_v$ that moves upward due to convection. The gradients of wind velocity and humidity can be used to calculate this flux (Verdon, 1993).

The ratio of the momentum flux to the vapor flow can be used to define the Thornthwaite-Holzman equation for vapor transport. The difference between the vapor pressure at the surface e_{as} and the vapor pressure at a height z_2 above the water surface e_a is used by the aerodynamic method's evaporation equation to get the evaporation rate E_a. The vapor pressure at the surface is the saturation vapor pressure at ambient air temperature, while the vapor pressure at height z2 is considered the ambient vapor pressure in the air (Giles & Drela, 1987).

$$E_a = B(e_{as} - e_a) \tag{17}$$

The units of E_a are mm/day and B represents the vapor transfer coefficient, measured in mm/day Pa. It is expressed as:

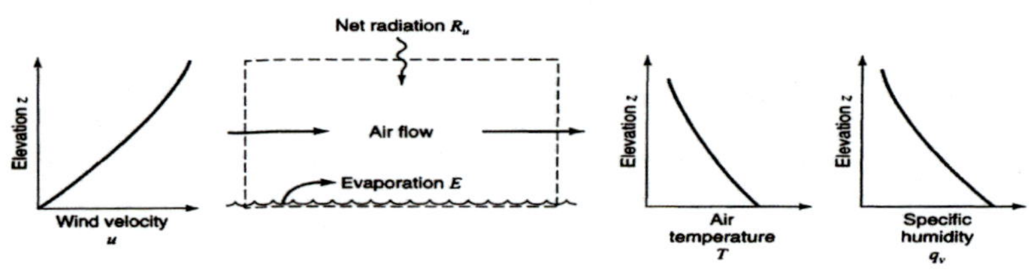

Figure 3.12. *Evaporation from an open water surface.*

Source: Woodward, D.; Creative Commons License.

$$B = \frac{0.102u_2}{[\ln(z_2/z_0)]^2} \tag{18}$$

When is considered as the wind velocity (m/s), it is important to take into account the measurement height z_2 (cm) and z_0 roughness height (0.01–0.06 cm) of the water surface. The units of vapor pressures are in Pa (N/) (Boggs, 1992). An approximation is used to calculate the saturation vapor pressure (Prueger & Kustas, 2005).

$$e_{as} = 611 \exp\left(\frac{17.27T}{237.3+T}\right) \tag{19}$$

and the vapor pressure is:

$$e_a = R_h e_{as} \tag{20}$$

where; R_h is the relative humidity ($R = e_a / e_s$ and ($0 \leq R_h \leq 1$); and T is the air temperature in degree Celsius (°C).

3.4. PERCOLATION AND GROUNDWATER

Learning Objectives:

- Understand groundwater dynamics, including recharge, water table, and aquifers.

Non-evaporated water from the soil may penetrate downwards through the unsaturated zone, often referred to as the vadose zone, and ultimately reach the groundwater zone. The groundwater zone refers to the region where all the pores are saturated with water, or with bedrock, where any fissures in the rock are filled with water. Diffuse recharging of groundwater refers to the downward movement of water from the surface (Berkowitz & Balberg, 1993).

The water table, which is the upper boundary of the groundwater zone, can vary in its proximity to the surface. It may either be relatively close or located at a significant depth. After becoming part of the groundwater, the water droplet will shift horizontally due to variations in the hydraulic head, which refers to differences in the water table's elevation (Marino, 1974).

Understanding the properties of a groundwater-containing formation is crucial. One key aspect is the amount of water it holds, known as porosity. Another important factor is how easily the water can move through the formation, which is determined by hydraulic conductivity. Larger pores, like those found in areas with a subsurface made of sand and gravel, have higher conductivities. This means that water can move more easily through these formations, which are known as aquifers. As a result, pumping out water from these aquifers is relatively simple (Poulain et al., 2018).

There are several ways in which surface water and groundwater can interact. At a stream, lake, or wetland, groundwater rises to the surface. Depending on the situation, the flow direction is either from surface water to groundwater (discharge of groundwater to surface water) or from surface water to groundwater (localized recharge of groundwater from surface water) (Xu et al., 2019).

SUMMARY

- The "Hydrology Principles and Processes" chapter offers a thorough summary of the key ideas in hydrology. The chapter begins with an examination of the foundations, including the definition of hydrology and the hydrologic cycle. It then explores the complexities of hydrologic systems and how ocean and atmospheric circulation affect the distribution of water on Earth.
- The chapter describes precipitation, including how it forms, what kinds of it there are, how it varies, and how rainfall is disposed of in watersheds. After that, a thorough discussion of evaporation is given, with an emphasis on the energy balance and aerodynamic methods for calculating evaporation rates.
- The chapter concludes with a discussion of groundwater and percolation, showing how water moves between the saturated and unsaturated zones of the Earth's subsurface.
- Overall, "Hydrology Principles and Processes" provides a solid foundation in the study of hydrology and sheds light on the distribution, mobility, and management of Earth's valuable water resources.

REVIEW QUESTIONS

1. What is hydrology, and why is it important for understanding Earth's water systems?
2. Describe the main processes involved in the hydrologic cycle.
3. How does precipitation variability impact hydrological systems?
4. Explain the factors influencing evaporation and the methods used to estimate evaporation rates.
5. Discuss the role of percolation in groundwater recharge and movement.
6. How do atmospheric and ocean circulation patterns influence regional hydrology?

MULTIPLE CHOICE QUESTIONS

1. Which of the following best defines hydrology?

a. The study of water movement in oceans

b. The study of water movement in rivers only

c. The study of water movement and distribution on Earth

d. The study of water movement in glaciers

2. The hydrologic cycle primarily involves which of the following processes?

a. Condensation, evaporation, and precipitation

b. Sedimentation, erosion, and runoff

c. Freezing, melting, and sublimation

d. Tidal action, wave erosion, and river meandering

3. Which of the following is an example of a hydrologic system?

a. A single raindrop

b. A lake and its watershed

c. A cloud formation

d. A mountain range

4. Atmospheric and ocean circulation plays a significant role in:

a. Weather patterns

b. Soil composition

c. Plant growth

d. Animal migration

5. What are the main types of precipitation?

a. Rain, hail, and fog

b. Snow, sleet, and dew

c. Frost, mist, and drizzle

d. Storm, hurricanes, and tornado

Answers to Multiple Questions

1. (c); 2. (a); 3. (b); 4. (a); 5. (b)

REFERENCES

1. Berkowitz, B., & Balberg, I. (1993). Percolation theory and its application to groundwater hydrology. *Water Resources Research, 29*(4), 775–794. https://doi.org/10.1029/92WR02707.

2. Boggs, D. W. (1992). Validation of the aerodynamic model method. *Journal of Wind Engineering and Industrial Aerodynamics, 42*(1–3), 1011–1022. https://doi.org/10.1016/0167-6105(92)90122-R.

3. Bourgouin, P. (2000). A method to determine precipitation types. *Weather and Forecasting, 15*(5), 583–592. https://doi.org/10.1175/1520-0434(2000)015<0583>2.0.CO;2.
4. Bryan, K., Manabe, S., & Pacanowski, R. C. (1975). A global ocean-atmosphere climate model. Part II. The oceanic circulation. *Journal of Physical Oceanography, 5*(1), 30–46. https://doi.org/10.1175/1520-0485(1975)005<0030>2.0.CO;2.
5. Burt, C. M., Mutziger, A. J., Allen, R. G., & Howell, T. A. (2005). Evaporation research: Review and interpretation. *Journal of Irrigation and Drainage Engineering, 131*(1), 37–58. https://doi.org/10.1061/(ASCE)0733-9437(2005)131:1(37).
6. Chen, Y., Ebert, E. E., Walsh, K. J., & Davidson, N. E. (2013). Evaluation of TRMM 3B42 precipitation estimates of tropical cyclone rainfall using PACRAIN data. *Journal of Geophysical Research: Atmospheres, 118*(5), 2184–2196. https://doi.org/10.1002/jgrd.50250.
7. Edwards, P. J., Williard, K. W., & Schoonover, J. E. (2015). Fundamentals of watershed hydrology. *Journal of Contemporary Water Research & Education, 154*(1), 3–20. https://doi.org/10.1111/j.1936-704X.2015.03214.x.
8. Frumin, G. T., Ivanov, M. E., Kulikova, L. A., & Eremina, A. V. (2017). Approaches to rainfall estimation and forecasting for urban wastewater disposal. *Proceedings of the Estonian Academy of Sciences, 66*(1), 29–99. https://doi.org/10.3176/proc.2017.1.04.
9. Giles, M. B., & Drela, M. (1987). Two-dimensional transonic aerodynamic design method. *AIAA Journal, 25*(9), 1199–1206. https://doi.org/10.2514/3.9752.
10. Horton, R. E. (1933). The role of infiltration in the hydrologic cycle. *Eos, Transactions American Geophysical Union, 14*(1), 446–460. https://doi.org/10.1029/TR014i001p00446.
11. Irmak, S., Skaggs, K. E., & Chatterjee, S. (2014). A review of the Bowen ratio surface energy balance method for quantifying evapotranspiration and other energy fluxes. *Transactions of the ASABE, 57*(6), 1657–1674. https://doi.org/10.13031/trans.57.10338.
12. Jansen, M. F. (2017). Glacial ocean circulation and stratification are explained by reduced atmospheric temperature. *Proceedings of the National Academy of Sciences, 114*(1), 45–50. https://doi.org/10.1073/pnas.1610438113.
13. Javanmard, S., Yatagai, A., Nodzu, M. I., BodaghJamali, J., & Kawamoto, H. (2010). Comparing high-resolution gridded precipitation data with

satellite rainfall estimates of TRMM_3B42 over Iran. *Advances in Geosciences, 25*, 119–125. https://doi.org/10.5194/adgeo-25-119-2010.

14. Jones, J. H., Olsen, F. J., Patterson, J. J., & Rushing, K. A. (1984). The pollution potential of rainfall runoff from a dairy liquid waste disposal site. *Agricultural Wastes, 10*(3), 177–186. https://doi.org/10.1016/0141-4607(84)90043-4.
15. Klots, C. E. (1991). Systematics of evaporation. *Zeitschrift für Physik D Atoms, Molecules and Clusters, 20*, 105–109. https://doi.org/10.1007/BF01383974.
16. Kousky, V. E., Kagano, M. T., & Cavalcanti, I. F. (1984). A review of the Southern Oscillation: Oceanic-atmospheric circulation changes and related rainfall anomalies. *Tellus A, 36*(5), 490–504. https://doi.org/10.1111/j.1600-0870.1984.tb00264.x.
17. Le Barbé, L., Lebel, T., & Tapsoba, D. (2002). Rainfall variability in West Africa during the years 1950–90. *Journal of Climate, 15*(2), 187–202. https://doi.org/10.1175/1520-0442(2002)015<0187>2.0.CO;2.
18. Lee, J. H., & Heo, J. H. (2011). Evaluation of estimation methods for rainfall erosivity based on annual precipitation in Korea. *Journal of Hydrology, 409*(1–2), 30–48. https://doi.org/10.1016/j.jhydrol.2011.07.031.
19. Leray, S., Engdahl, N. B., Massoudieh, A., Bresciani, E., & McCallum, J. (2016). Residence time distributions for hydrologic systems: Mechanistic foundations and steady-state analytical solutions. *Journal of Hydrology, 543*, 67–87. https://doi.org/10.1016/j.jhydrol.2016.04.066.
20. Linton, J. (2008). Is the hydrologic cycle sustainable? A historical-geographical critique of a modern concept. *Annals of the Association of American Geographers, 98*(3), 630–649. https://doi.org/10.1080/00045600802046619.
21. Loaiciga, H. A., Valdes, J. B., Vogel, R., Garvey, J., & Schwarz, H. (1996). Global warming and the hydrologic cycle. *Journal of Hydrology, 174*(1–2), 83–127. https://doi.org/10.1016/0022-1694(95)02753-1.
22. Manabe, S. (1969). Climate and the ocean circulation: I. The atmospheric circulation and the hydrology of the earth's surface. *Monthly Weather Review, 97*(11), 739–774. https://doi.org/10.1175/1520-0493(1969)097<0739>2.3.CO;2.
23. Manabe, S., Bryan, K., & Spelman, M. J. (1975). A global ocean-atmosphere climate model. Part I. The atmospheric circulation. *Journal of Physical Oceanography, 5*(1), 3–29. https://doi.org/10.1175/1520-0485(1975)005<0003>2.0.CO;2.

24. Marino, M. A. (1974). Growth and decay of groundwater mounds induced by percolation. *Journal of Hydrology, 22*(3–4), 295–301. https://doi.org/10.1016/0022-1694(74)90020-1.

25. Mehdipour, I., Ganji, D. D., & Mozaffari, M. (2010). Application of the energy balance method to nonlinear vibrating equations. *Current Applied Physics, 10*(1), 104–112. https://doi.org/10.1016/j.cap.2009.07.003.

26. Nesbitt, S. W., Cifelli, R., & Rutledge, S. A. (2006). Storm morphology and rainfall characteristics of TRMM precipitation features. *Monthly Weather Review, 134*(10), 2702–2721. https://doi.org/10.1175/MWR3200.1.

27. O'Connell, P. E., & Todini, E. (1996). Modeling of rainfall, flow and mass transport in hydrological systems: An overview. *Journal of Hydrology, 175*(1–4), 3–16. https://doi.org/10.1016/S0022-1694(96)80003-0.

28. Odening, M., Mußhoff, O., & Xu, W. (2007). Analysis of rainfall derivatives using daily precipitation models: Opportunities and pitfalls. *Agricultural Finance Review, 67*(1), 135–156. https://doi.org/10.1108/00214660780001222.

29. Penman, H. L. (1956). Estimating evaporation. *Eos, Transactions American Geophysical Union, 37*(1), 43–50. https://doi.org/10.1029/TR037i001p00043.

30. Poulain, A., Watlet, A., Kaufmann, O., Van Camp, M., Jourde, H., Mazzilli, N., & Hallet, V. (2018). Assessment of groundwater recharge processes through karst vadose zone by cave percolation monitoring. *Hydrological Processes, 32*(13), 2069–2083. https://doi.org/10.1002/hyp.13135.

31. Prueger, J. H., & Kustas, W. P. (2005). Aerodynamic methods for estimating turbulent fluxes. In J. L. Hatfield (Ed.), *Micrometeorology in Agricultural Systems* (pp. 407–436). https://doi.org/10.2134/agronmonogr47.c18.

32. Rosinski, J. (1974). Role of aerosol particles in the formation of precipitation. *Reviews of Geophysics, 12*(1), 129–134. https://doi.org/10.1029/RG012i001p00129.

33. Salas, J. D., Govindaraju, R. S., Anderson, M., Arabi, M., Francés, F., Suarez, W., & Green, T. R. (2014). Introduction to hydrology. In H. V. Gupta (Ed.), *Modern Water Resources Engineering* (pp. 1–126). https://doi.org/10.1007/978-3-319-07548-8_1.

34. Shahid, S. (2010). Rainfall variability and the trends of wet and dry periods in Bangladesh. *International Journal of Climatology, 30*(15), 2299–2313. https://doi.org/10.1002/joc.2053.

35. Shaw, S. B., Beslity, J. O., & Colvin, M. E. (2019). Working towards a more holistic set of hydrologic principles to teach nonhydrologists: Five

simple concepts within catchment hydrology. *Hydrological Processes, 33*(16), 1–99. https://doi.org/10.1002/hyp.13412.

36. Smith, R. S., Dubois, C., & Marotzke, J. (2006). Global climate and ocean circulation on an aquaplanet ocean-atmosphere general circulation model. *Journal of Climate, 19*(18), 4719–4737. https://doi.org/10.1175/JCLI3895.1.
37. Spittlehouse, D. L., & Black, T. A. (1980). Evaluation of the Bowen ratio/energy balance method for determining forest evapotranspiration. *Atmosphere-Ocean, 18*(2), 98–116. https://doi.org/10.1080/07055900.1980.9649078.
38. Stocker, T. F., & Wright, D. G. (1996). Rapid changes in ocean circulation and atmospheric radiocarbon. *Paleoceanography, 11*(6), 773–795. https://doi.org/10.1029/96PA02640.
39. Sun, Q., Miao, C., Duan, Q., Ashouri, H., Sorooshian, S., & Hsu, K. L. (2018). A review of global precipitation data sets: Data sources, estimation, and intercomparisons. *Reviews of Geophysics, 56*(1), 79–107. https://doi.org/10.1002/2017RG000574.
40. Tanner, C. B., & Pelton, W. L. (1960). Potential evapotranspiration estimates by the approximate energy balance method of Penman. *Journal of Geophysical Research, 65*(10), 3391–3413. https://doi.org/10.1029/JZ065i010p03391.
41. Trenberth, K. E., Dai, A., Rasmussen, R. M., & Parsons, D. B. (2003). The changing character of precipitation. *Bulletin of the American Meteorological Society, 84*(9), 1205–1218. https://doi.org/10.1175/BAMS-84-9-1205.
42. Verdon, J. M. (1993). Review of unsteady aerodynamic methods for turbomachinery aeroelastic and aeroacoustic applications. *AIAA Journal, 31*(2), 235–250. https://doi.org/10.2514/3.11353.
43. Volford, A., Izsák, F., Ripszám, M., & Lagzi, I. (2007). Pattern formation and self-organization in a simple precipitation system. *Langmuir, 23*(3), 961–964. https://doi.org/10.1021/la061557g.
44. Wagener, T., Sivapalan, M., Troch, P. A., McGlynn, B. L., Harman, C. J., Gupta, H. V., & Wilson, J. S. (2010). The future of hydrology: An evolving science for a changing world. *Water Resources Research, 46*(5), 1–99. https://doi.org/10.1029/2009WR008906.
45. Winton, M. (2003). On the climatic impact of ocean circulation. *Journal of Climate, 16*(17), 2875–2889. https://doi.org/10.1175/1520-0442(2003)016<2875>2.0.CO;2.

46. Xu, B., Shao, D., Fang, L., Yang, X., Chen, S., & Gu, W. (2019). Modeling percolation and lateral seepage in a paddy field-bund landscape with a shallow groundwater table. *Agricultural Water Management, 214*, 87–96. https://doi.org/10.1016/j.agwat.2019.01.010.

CHAPTER 4

FLOOD RISK ASSESSMENT AND MANAGEMENT

LEARNING OBJECTIVES

At the end of this chapter, readers will be able to:

1. Understand flood risk components: hazard, vulnerability, consequence, and risk.
2. Understand flood risk management: pre-flood, event, and post-flood phases.
3. Recognize flood risk tasks: analysis, assessment, and reduction.
4. Recognize measures and instruments for flood risk reduction.
5. Understand the SPRC model and its application in flood risk assessment.
6. Understand the interconnected tasks of flood risk management.
7. Explore the societal influences on flood risk decision-making.
8. Recognize challenges and strategies for integrated flood risk management.

INTRODUCTORY EXAMPLE

Riverine Solutions Inc., a consulting firm specializing in flood risk assessment and management, tasked with developing a comprehensive flood risk management plan for a municipality situated along a major river. This municipality has experienced recurrent flooding, causing significant damage to infrastructure, homes, and businesses. In response to these challenges, the local government has commissioned Riverine Solutions Inc. to assess flood risks and propose strategies to mitigate them.

Using advanced modeling techniques and data analysis, Riverine Solutions Inc. conducts a thorough assessment of the area's flood hazards, considering factors such as historical flood patterns, topography, land use, and climate projections. They identify high-risk areas prone to flooding and assess the vulnerability of critical infrastructure, such as hospitals, schools, and transportation networks.

Based on their findings, Riverine Solutions Inc. develops a tailored flood risk management plan, which includes a combination of structural and non-structural measures. Structural measures may involve constructing levees, floodwalls, and storm water drainage systems to protect vulnerable areas from inundation. Non-structural measures may include land-use zoning regulations, community education programs, and early warning systems to enhance preparedness and resilience.

Throughout the planning process, Riverine Solutions Inc. collaborates closely with local stakeholders, including government agencies, community groups, and residents, to ensure that the proposed strategies align with their needs and priorities. By leveraging their expertise and engaging stakeholders in decision-making, Riverine Solutions Inc. aims to develop a robust flood risk management plan that reduces the municipality's vulnerability to future flooding events.

UNIT INTRODUCTION

Floods pose a significant danger to human societies. It is clear from the rise in damages over the past 50 years caused by a succession of severe floods. In recent times, a tsunami that hit Southeast Asia resulted in a staggering loss of 2,20,000 lives, making it among the most devastating floods in recent history (Vorogushyn et al., 2018). Throughout the International Decade of Natural Disaster Reduction (IDNDR) from 1990 to 1999, there was a growing recognition that the traditional approach of "flood protection" was not suitable. Attaining complete protection is simply not feasible or practical due to the exorbitant expenses involved and the inherent uncertainty of the situation. Instead, it has been suggested that risk management is a more acceptable approach, and this perspective is gaining increasing recognition in flood research (Merz et al., 2014). Furthermore, there is a noticeable shift in various countries' environmental and regional policies, moving away from solely focusing on flood protection and towards a more comprehensive approach to flood risk management (Solin & Skubincan, 2013).

Did You Know?

Did you know that the largest flood ever recorded took place in China in 1931? The event went on for a few months and had a significant impact on approximately 51 million individuals, resulting in extensive destruction and loss of life.

Flood risk management covers a wide range of problems and tasks, such as forecasting flood hazards, assessing their social effects, and taking measures to mitigate risk. Managing flood risks effectively necessitates a comprehensive and cohesive approach, considering the multitude of factors involved (Gouldby et al., 2008). This chapter describes the important terms, an extensive description of components and tasks, as well as an essential framework for effectively managing flood risks. Given these factors, there are challenges for both academic research and real-world implementation that arise mainly from embracing a holistic approach focused on evaluating and controlling risks (Lyu et al., 2019).

4.1. TERMS AND CONCEPTS

Learning Objectives:

- Understand flood risk components: hazard, vulnerability, and consequences.
- Understand flood risk management modes: pre-flood, event, post-flood.

4.1.1. Flood Risk

Floods are characterized as a temporary inundation of land by water beyond its usual boundaries. They occur in several water bodies, including both large and small basins of rivers, estuaries, coastal areas, and local regions. In addition to these general circumstances, floods can be categorized based on the events that cause them. These include winter rainfall, summer convectional storms, snowmelt, sea surges, tides, tsunamis, and caused by rising groundwater, urban sewer floods, and resulting from dam breaks or reservoir control. Flash floods are a unique and rapidly changing form of flooding (Hall et al., 2003). Flood events can be described by various factors, including matter fluxes, water depth, temporal and spatial dynamics, and flow velocity. In many instances, flooding is a natural occurrence that may not be considered a danger in areas like natural floodplains. However, floods in heavily utilized catchments are frequently impacted by human activities such as land use and river training (Baan & Klijn, 2004).

Flood hazard refers to the likelihood of possibly catastrophic flood events. There is a possibility of damage when elements are exposed to floods, but it is not guaranteed. There are different characteristics associated with flood hazards. As an example, a building located in a floodplain may face potential threats from a 50-year flood, where the water level reaches 1 meter, and a 100-year flood, where the water level rises to 1.5 meters. In addition, these occurrences can be linked to various transportation abilities in terms of debris, sediment, and other potentially harmful substances, each with different effects on humans and the environment (Merz et al., 2010).

The extent of damage caused by flood hazards is influenced by the vulnerability of the elements that are exposed to them. Vulnerability is a concept that refers to the inherent characteristics of certain factors, which assess how likely they are to be harmed. The concept can be seen as a blend of vulnerability and societal importance, manifested through both direct and indirect impacts, whether they are intangible or not. Unlike social value, which is unaffected by the hazard, susceptibility refers to the method of damage generation (Diaconu et al., 2021). The outcome is influenced by the specific characteristics of the flood event and the vulnerability of those components to danger (Shah et al., 2018). There are three main areas to consider when it comes to flood vulnerability: social and cultural, economic, and ecological vulnerability. These areas are crucial for ensuring sustainability. Social or cultural vulnerability encompasses a range of consequences, including the potential for loss of life, medical impacts such as injuries, diminished well-being, increased stress, social disruptions, loss of personal belongings, and the erosion of cultural heritage. Economic vulnerability refers to the financial losses incurred due to harm caused by property assets, goods, reduced productivity, basic materials, and the need for relief efforts (Lechowska, 2022).

The vulnerability of ecosystems includes the contamination of water, soil, and ecological systems by human activities and their living organisms. Flood risk arises from the interaction of danger of flooding and vulnerability to floods. Flood risk assessment and management in academia involves evaluating the likelihood of adverse outcomes resulting from floods. This assessment is contingent upon the vulnerability of elements that are susceptible to potential flood hazards. The concept of risk has evolved since the early days of risk research (Burningham et al., 2008). When comes to floods, it is seen as damage to elements that are prone to flooding and have a particular vulnerability ("elements at risk") because of potential flood events and their characteristics. It is important to note that flood risk assessment and management should not be mistaken for reliability risk, which is crucial for evaluating the security of structural measures for flood protection (Plate, 2002).

A conceptual model known as the Source-Pathway-Receptor-Consequence Model (SPRC Model) was put forward to explain flood risk. Please refer to Figure 4.1 for a visual representation. The process illustrates a straightforward sequence starting with meteorological and hydrological events, which may occur inland or along the coasts. These events then lead to discharge and inundation, which in turn have physical impacts on elements at risk. Finally, the effects of these impacts are assessed. The connection between 'source,' 'pathway,' and 'receptor' pertains to the physical procedure, while evaluating the 'negative consequence' is influenced by societal values (Morrison et al., 2018).

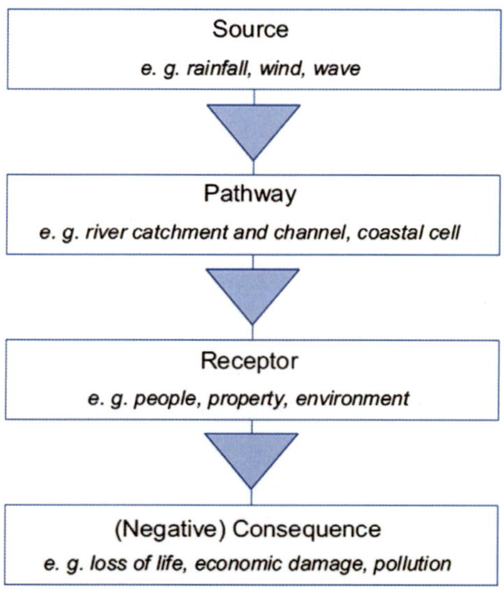

Figure 4.1. *Illustration source-pathway-receptor-consequence-model.*

Source: Faulkner; Creative Commons License.

When it comes to flood risk, the "source" and "pathway" are crucial factors that contribute to the flood hazard. The determination of 'source' is based on the probability (p) of floods with a specific magnitude and other characteristics (m). Two important factors in reducing the risk of inland floods are early warning

(w) systems and the ability of source areas to retain water (t). The 'pathway' can be characterized by either inland discharge or coastal overflow and inundation, along with different attributes (a) and interventions for flood control (c). 'Receptor' and '(negative) consequence' indicate the vulnerability, while 'receptor' specifies the susceptibility with interventions to enhance resistance and resilience (r). 'Consequence' refers to the negative impact on values, such as (v; damage), that can be mitigated or compensated through interventions (d) (De Bruijn, 2004). Thus, the expression of flood risk is represented by this function:

$$Flood\ risk = f((p,m,w,t))_{source}, (i,a,c)_{pathway}, (s,r)_{receptor}, (v,d)_{consequences})$$

For every component at risk and flood hazard, the SPRC model's causal chain exists. In addition, there are intricate connections between different pathways, interventions for flood control, and the vulnerability of at-risk elements. There are instances where the interrelations involve multiple suggestions (Gersonius et al., 2008). Here a system is defined that encompasses all relevant elements and processes as a "flood risk system." When it comes to inland floods, the focus is on river catchments, while coastal floods are concerned with coastal cells, areas that are hydraulically connected. The overall risk of a flood risk system can be characterized as the cumulative risk of all its components (Klijn et al., 2015).

4.1.2. Flood Risk Management

In the literature on floods, the word management is utilized in at least two distinct contexts, with some sources excluding risk analysis while others include it. One aspect is the reliability of the current flood defense structures from a hydrological perspective. Management is regarded as the process of making decisions and taking actions to reduce the risk that remains above the standards set for flood protection. In the past, the assessment of remaining risk has been conducted via scientific investigations (Thaler & Hartmann, 2016). Addressing flood risks within this scenario involves analyzing potential flood risks and subsequently implementing measures to manage and mitigate these risks. Another interpretation characterizes management as the process of making informed choices and implementing strategies to analyze, evaluate, and mitigate the potential dangers posed by floods. In this scenario, flood risk management encompasses the analysis, assessment, and reduction of risks associated with flooding. Both concepts are distinct options and are not easily merged (Hartmann & Driessen, 2017).

The concept of "management" originated in the field of business economics. It refers to the various activities that effectively and efficiently regulate the decisions and actions taken by an actor, organization, or network. These activities involve various tasks such as data collection, analysis, goal setting, evaluating options, and more. They also include organizing, monitoring, controlling, staffing, directing, and learning. Within the context of European Water Policy, the term "management" is already employed. An example is given by the specifications of River Basin Management Plans defined in Annex VII in the European Water Framework Directive. In addition to other aspects, they cover the examination of pressures and impacts, the evaluation of the water status, and a set of measures (Waghwala & Agnihotri, 2019).

Given the circumstances, it is advised to describe flood risk management as a comprehensive and ongoing analysis, assessment, and decrease in flood risk that considers the needs of society. The risk of flood systems should be approached from a comprehensive perspective, taking into account all relevant factors. The concept of "continuous" emphasizes the importance of consistently evaluating flood risks, monitoring their ever-changing nature, and assessing the impact of mitigation efforts (Thumerer et al., 2000). Analysis, assessment, and interventions for decreasing risks are essential components of development models that focus on the responsible managing entity. When it comes to the oversight of the risk of flood systems, society, including politicians, experts, and individuals, can be viewed as the key players responsible for overseeing it. These experts evaluate the potential for flooding and determine whether a particular flood risk is acceptable or not. Then make decisions on how to minimize the risk through various interventions (Marchi et al., 2010).

Flood risk management occurs through a collaborative process involving various stakeholders. Based on possible changes in the flood risk system, these actors come from adjacent areas (including multiple municipalities), different fields (such as water authority and spatial planning authority), and various levels (ranging from local to international). The decision-making and development process can vary greatly based on the different planning, political, administrative, and cultural systems in place (Cvetkovic & Martinović, 2020).

Throughout the ongoing management procedure, various modes of management can be identified: the flood event, the pre-flood, and the post-flood modes. These concepts have been used to organize and structure activities aimed at reducing the risks associated with floods. Based on an extensive knowledge of flood risk management, the approaches also encompass risk analysis and risk assessment. For risk analyses, it is already common to differentiate between these modes as long-term' and 'operational' approaches, each with their methodological requirements (Tran et al., 2008).

The focus of the pre-flood manner is to successfully reduce flood risks over the long term. It is often defined by the presence of enough time and resources. A thorough examination and precise analysis can be conducted to study flood risk. Various approaches to proactive or precautionary measures may be extensively debated, even within less formal institutional frameworks. On the other hand, the way the flood event unfolds is affected by the event management mode (Foudi et al., 2015). Response times are brief and the available resources for preparing actions are limited when compared to the phase before the flood. Estimating the ever-changing direction of flood risk is crucial in the short term. Urgent and decisive actions must be promptly implemented and properly documented. The focus after a flood involves the recovery process and preventing any additional negative outcomes. The temporal dynamic and organizational structure can be categorized as transitioning from event management to the pre-flood mode (Birkholz et al., 2014).

According to certain authors, there is a cyclical process in managing flood risks. It begins during a flood event, accompanied by the post-flood phase, and then proceeds to the pre-flood phase. Public awareness can be enhanced by using an image of a circle, which symbolizes the repetitive nature of floods and the need for risk reduction. It is important to keep in mind that past actions have an impact on future decisions in the field of research and flood risk management. The modes, therefore, adhere to a more accurate helical pattern. In addition, the different modes can sometimes intersect in real-world situations, such as when preventive measures are incorporated into recovery activities (Hartmann & Spit, 2016).

4.2. TASKS AND COMPONENTS

Learning Objectives:

- Comprehend flood risk management tasks: analysis, assessment, reduction.
- Understand components of flood risk analysis: hazard, vulnerability, risk.
- Know about risk assessment: perception, weighing, ALARP principle.
- Explore flood risk reduction measures: structural, non-structural, regulatory instruments.

According to flood risk management, there exist three tasks using specific components that can be utilized to structure the management activities (*refer* to Figure 4.2) (Stapleton & Halgren, 1987). Here are the primary responsibilities:

- Risk analysis;
- Risk assessment;

- Risk reduction.

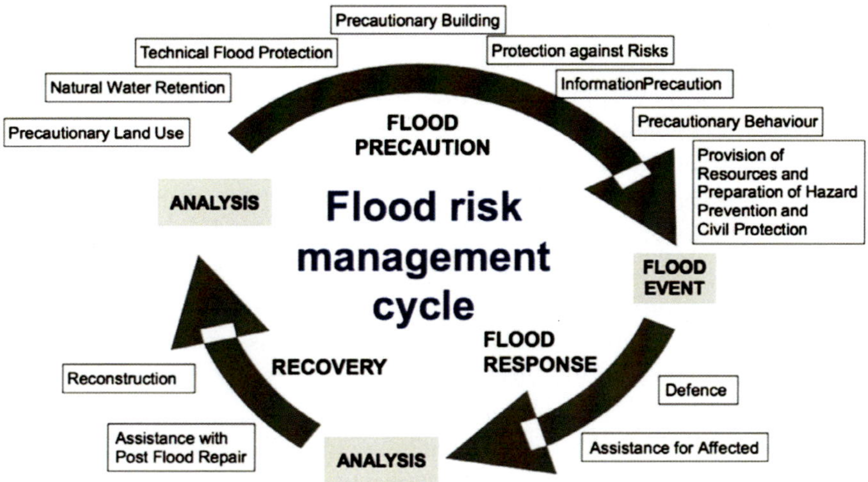

Figure 4.2. *Illustration of flood risk management cycle.*

Source: Cassel & Hinsberger; Creative Commons License.

Understanding the potential for floods, evaluating their impact, and implementing measures to minimize risks are all important aspects of flood management. To accomplish the objectives of each task, specific components are necessary. The topics covered include identifying potential risks and developing strategies for responding to floods. Various systematizations have been proposed in the research, but none of them align with the comprehensive and actor-oriented approach to understanding flood risk management presented in this context (Kray & Lindenberger, 2002) (Figure 4.3).

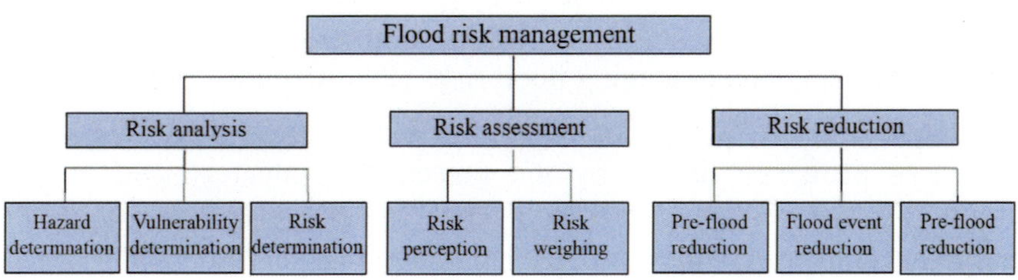

Figure 4.3. *Illustration of the tasks and components of flood risk management.*

Source: Gareth Pender; Creative Commons License.

4.2.1. Risk Analysis

Flood risk analysis determines risks associated with past, present, or future events, such as proposed activities or global change. The assessment involves evaluating the potential for flooding, the susceptibility to flood damage, and the risk posed by floods. Various approaches from both natural and social sciences are necessary to explain the flood risk system. It is important to incorporate various methods, such as meteorological, economic, social science, hydrological, hydraulic, and ecological approaches when using the SPRC-Model. One of the main challenges is finding a way to collect and combine knowledge from various disciplines. Generally, there is already a wealth of experience in combining deterministic modeling with statistical evaluation of hydro-meteorological procedures. Real probabilistic methods are scarce (Mackenzie et al., 1992). In terms of the socio-economic field, certain approaches focus on addressing the impact of economic losses. Understanding and approaches to indirect consequences and impacts on society and the environment are still in the early stages of development. It is quite uncommon to find comprehensive approaches that attempt to integrate multiple fields. Lately, there has been a growing focus on the development of intricate model systems. In addition, there are tiered methodologies available for conducting risk analysis at different levels (Akintoye & MacLeod, 1997).

Although there have been efforts to create more integrated approaches, it is important to acknowledge that our understanding of flood risk systems and the effects of reducing flood risk interventions will always be limited. Uncertainty is a crucial factor to consider when analyzing flood risk. Two types of uncertainty can be distinguished: aleatory and epistemic uncertainty. Random uncertainty is a result of the inherent variability of quantities across different periods, locations, or individuals. Uncertainty arises from the limited understanding of the components and procedures involved in the risk of flood systems. It becomes more evident when considering that risk analysis often focuses solely on aspects that can be easily quantified. The narrow approach to constructing heuristics inevitably results in a limited perspective of the actual system for assessing and managing flood risks (Paté-Cornell, 1996).

Essentially, an awareness of flood risk management to be an ongoing procedure is primarily driven by the presence of aleatory and epistemic uncertainty. It is necessary to regularly repeat the risk evaluation of the flood system to account for changes in the system and evaluate the effectiveness of risk reduction measures. However, it is important to recognize that uncertainty is not something that can be completely avoided. Various techniques exist for its measurement and evaluation. These methods provide a way to consider the level of uncertainty, which can be observed as valuable supplementary information for managing flood risks (Fishburn, 1977). Currently, flood risk analysis yields two main outcomes: risk maps in both printed and digital formats, and operational procedures such as long-term flood forecasting and real-time flood warning systems. Maps that assess the risk of flooding condense the complex flood risk system into a simplified, two-dimensional format. These maps offer valuable insights into the likelihood of flooding, water levels, flow velocity, sediment transport, and more. Various approaches are commonly used. Their dependability relies on the precision and requirements of the remaining uncertainty in the risk analysis. There are already some risk map systems available that are more flexible, interactive, and web-based. Additionally, further development is underway to enhance these systems (Short, 1984). Current flood forecasting and warning systems offer valuable data regarding active flood situations. Their goal is to improve the amount of time available for preparatory activities such as reservoir control and evacuation. The focus is primarily on meteorological and hydrological modeling. Comprehensive risk-based approaches, including detailed information on potential damage, have not yet been made available. Typically, vulnerability is assessed based on critical peak discharges rather than anticipated damage. However, it is widely recognized that the current systems for managing flood events are of great importance.

The way society perceives and weighs the tolerability of different flood risks can greatly influence the interpretation of scientific analyses. Does applying this distinction imply that one should prioritize a more precise and objective risk analysis over a less valid subjective assessment? On the one hand, it recognizes the importance of understanding the threat of floods through theories and methods. On the other hand, it acknowledges the significance of both collective and individual perception and how society weighs the risks. In the following, the concepts of risk perception and risk weighing are explored using a two-step approach to assess potential risks (Merz et al., 2014).

Perception of risk is influenced by the backgrounds and experiences of people and organizations involved in managing floods. For instance, individuals who have previously encountered a severe flood event are likely to have a distinct perspective on flood risks compared to others. There are different approaches to understanding risk perception, including psychological-cognitive and sociological-cultural perspectives. Psychological-cognitive approaches center on the various factors that impact our non-rational awareness and decision-making processes. They have a particular interest in understanding the importance of personal and shared values, emotions, experiences, and perspectives when it comes to perceiving actual risks. By comparing a rational and presumably accurate flood risk assessment with subjective awareness and decisions, these strategies align with the idea of using 'objective' data derived from scientific risk analysis as a foundation for societal risk perception. However, there is an ongoing critical discussion regarding the potential for accurately interpreting factors related to risk perception (Solin & Skubincan, 2013).

The sociological-cultural approaches encompass a diverse field. There is a clear distinction between research on risk communication and cultural anthropological searches. Research on risk communication emphasizes the importance of effective communication among experts and the public. It focuses on understanding the conditions, issues, and strategies for improving the transfer of knowledge. Recent research has further developed the role of communication in assessing and managing risks, shifting from a passive approach to actively engaging in the negotiation of risk acceptability. A sense of risk is believed to be influenced by the cultural context in cultural anthropology. The selection process over risk perception is attributed to the various cultural organization kinds within society (Lyu et al., 2019).

Due to the limitations of this chapter, it is not possible to provide an in-depth analysis of the different theories. In this context, the perception of risk is seen as the creation of a potential negative outcome through individual or collective imagination. The understanding of this concept is shaped by the values, emotions, personal encounters, and viewpoints that are shaped by the societal culture. The perception of risk does not develop gradually or about a specific, tangible threat (Prieto et al., 2020).

Society may not have a direct perception of the overall risk associated with flood risk systems. Through risk analyses, valuable information is obtained and contributes to a holistic understanding of risk perception. From this perspective, risk perception can be seen as a comprehensive framework for analyzing risks. Given the various paradigms involved, it appears that the proposed differentiation between risk analysis and risk perception is suitable. Risk analyses are grounded in scientific principles, while risk perception reflects real-world attitudes (Schröter et al., 2021).

Considering the perception of risk alone does not automatically involve decisions on how to address the risk. It highlights the importance of considering different behavioral choices when assessing risk. There is a close relationship between options that have the potential for negative consequences, known as risk, and options that have the potential for positive consequences, known as opportunities. It is generally acknowledged that it is not possible to eliminate risk. Otherwise, numerous opportunities go unnoticed. According to Wildavsky, the absence of risk poses the greatest risk of all. Considering risk involves determining an acceptable level of potential "costs" about specific "benefits" (Buchecker et al., 2013).

Luhmann acknowledges the concept of risk, but only if there are identifiable drawbacks that can be linked to a particular decision. To account for the potential negative outcomes associated with risk, it is important to differentiate between a "causal risk" and a "normative risk" when making decisions, as outlined in risk analysis. The concept of "normative risk" pertains to the acceptability of a quantified "causal risk" in a formal context. Without clear specifications of the potential negative consequences, a normative risk assessment would lack meaning. It works both ways. It is important to note that relying solely on causal risk is not enough when it comes to making informed decisions about risk management. This approach fails to consider the necessary prerequisites for effective decision-making.

The evaluation of risks involves understanding the connections between causes and effects and determining the level of risk that is acceptable. The main question is: What level of risk can be tolerated, considering the anticipated costs and benefits? This assessment should consider the various advantages associated with floods or areas prone to flooding. Floods have the potential to create habitats for a wide range of species and can also be valuable resources for irrigation, not limited to sub-tropical regions. Urban flood-prone areas have significant economic value, thanks to their access to waterways and navigation, as well as their strategic locations near city centers (Tsakiris, 2014).

In recent academic literature, there is a prevalent concept for risk assessment known as the ALARP Principle, which stands for 'as low as reasonably practicable.' According to the engineering principle, there is an underlying assumption of pre-existing vulnerability and limited capacity for risk reduction. Considering the information provided earlier, it is evident that a comprehensive assessment of potential risks should consider the associated costs as well as the potential benefits. 'Costs' encompass both the adverse consequences and the efforts involved in reducing risks (see below). It is important to consider more than just the reduction efforts when assessing damages, as suggested by the ALARP Principle. Instead, it is important to compare the overall costs with the positive consequences (opportunities) that arise (Zeleňáková, 2011).

Without a doubt, there are immeasurable costs, such as the tragic loss of lives or injuries that cannot be compared to any potential benefits. It is generally agreed that there will be a widely accepted area of costs that is prohibited. Similarly, this applies to an area that is considered normal, where there are minor risks that cannot be avoided. The negotiation of the "border area" between the prohibited and normal zones is of utmost importance in flood risk management (Apel et al., 2004).

While the primary focus of risk assessment seems to be on cost-benefit analysis, it is important to consider the broader concept of sustainability. This involves taking into account the various tangible and intangible costs, both direct and indirect, which necessitate the use of multi-criteria approaches. They enable the incorporation of a diverse set of criteria and fuzzy-linguistic scales of thresholds. Various GIS-based MADM methods can be utilized from different areas of water management and environmental research. However, these methods have not been fully utilized in the assessment of multiple flood risks. Assessing these risks can become more intricate when considering the inclusion of costs and benefits from other aspects of water policy, such as mandated flood risk management plans and integrated river basins (Li et al., 2012).

4.2.2. Risk Reduction

When risks are deemed unacceptable, appropriate measures and tools are implemented to mitigate the risk. Measures are described as interventions that involve direct physical actions, while instruments are interventions that indirectly lead to measures or influence human conduct. Historically, the focus has been on using structural approaches for flood defense interventions and non-structural approaches for other types of interventions (Mulholland & Christian, 1999). One possible explanation for this could be the influence of the conventional engineering approach to managing flood risks. From a societal perspective, it is important to delve deeper into the concept of non-structural measures to better understand the factors that drive interventions. On the other hand, the term 'instrument' is already widely used to describe activities in different areas of financial and legal policies (Mitchell & McGoldrick, 1996).

Did You Know?

Certain animal species have evolved specialized adaptations to effectively deal with the challenges posed by floods. As an illustration, specific types of fish can sense variations in the pressure of water and react to the threat of flooding by seeking shelter in deeper parts of the water or swimming upstream.

There are two types of measures: permanent and temporary. Permanent measures involve making direct physical changes to the flood risk system that result in long-lasting improvements. On the other hand, temporary measures are implemented specifically to decrease the risk during ongoing flood events. Long-term solutions involve implementing engineering projects to control flooding and implementing measures to enhance the ability of infrastructure and buildings to withstand and recover from floods. Additionally, they incorporate measures to enhance the ability to hold water, decrease the maximum flow rate, and minimize the loss of materials from the surrounding areas (such as suspended and bed load, debris) in the areas upstream and on the floodplains

(Wald et al., 1990). Provisional measures include the use of demountable flood protection such as sandbags and movable flood barriers. Additionally, there may be a need to relocate individuals, animals, or goods either on-site or off-site through evacuation. It is also important to secure objects and goods by, for example, locking gas valves. In addition, the process of repairing flood damages, such as removing toxic sediment buildup, is only a temporary solution due to time limitations (Benson et al., 2001).

Different types of instruments can be used for managing flood risks. These instruments can be categorized into regulatory, financial, and communication mechanisms. Regulatory instruments are interventions that utilize legal or other standard mechanisms and are derived from various policy sectors such as water management (e.g., flood prevention acts), spatial planning (e.g., legally binding land use plans), environmental protection, and nature conservation (e.g., Strategic Environmental Assessment). Various financial instruments can either support or restrict activities related to flood risk (Cho & Lee, 2006). For example, incentives may be provided to encourage the flood testing of built structures, while zoning of insurance can impose restrictions. Various tools are utilized to facilitate effective communication, such as the dissemination of information through media, brochures, and literature. Additionally, informal collaboration among relevant actors plays a crucial role in fostering cooperation. Furthermore, education, warning, preparation, and instructing are key components, exemplified by the implementation of flood warning systems. Various instruments are utilized to program, regulate, implement, and occasionally provide funding for the realization of measures (Catania et al., 1990).

Measures and instruments are compatible with specific coefficients of the SPRC model (refer to the information provided above): Retention, for example, focuses on the factor 't' of sources. Flood control, on the other hand, addresses the factor 'c' of pathways. The enhancement of resilience and constructive resistance is aimed at the factor 'r' of receptors, while compensation deals with the factor 'd' of consequences (Skeem & Monahan, 2011). There are different approaches to handling flood risks, which can be categorized as actions taken before, during, and after a flood occurs.

Pre-flood interventions focus on reducing the impact of floods and protecting vulnerable areas. This includes implementing measures to decrease the magnitude of floods and safeguarding existing vulnerable elements. Additionally, behavioral preparedness plays a crucial role in effectively dealing with potential flood events. Some examples include studying how water and matter are retained in the source areas of river floods, implementing spatial development strategies in flood-prone areas such as zoning and building construction to reduce risk, designing flood control measures like reservoirs, polders, and dykes for protection, and preparing people who are at risk for potential floods (Alexander, 2013).

Flood event management involves predicting and alerting people about ongoing floods, implementing flood control measures to manage water levels, utilizing flood protection structures for defense, and responding to emergencies through evacuation and rescue efforts. Some examples include the use of flood forecasting and warning systems, both official and unofficial, to alert people to potential flooding. Flood control measures, such as managing reservoirs and polders, are also employed to minimize the impact of floods. Additionally, risk mitigation strategies, like evacuation and rescue operations, are implemented to ensure the safety of affected individuals during emergencies (Rugman, 1976).

Post-flood interventions involve various measures aimed at aiding in the recovery and reconstruction process. One example is the insurance coverage for damage caused by floods. There is a wide variation in the effects, effectiveness, and efficiency of different measures and instruments. The effects can vary greatly, both in their intended outcomes and unintended consequences, and are heavily influenced by the specific characteristics of the risk of flood system. Effectiveness is a measure of how successfully desired outcomes are achieved (Weichselgartner & Pigeon, 2015). The efficacy of measures and instruments is determined by the balance between their effectiveness and the efforts or resources they require. For example, the cost-efficiency of measures to prevent flooding refers to the relationship between the reduction in potential damage and the associated costs of implementing those measures. In this case, it requires details about capital investments or opportunity costs. The overall effectiveness of flood risk assessment and management relies on the functionality of various measures and mechanisms, as well as their impacts on the flood risk system. Additionally, both the monetary and non-monetary efforts put into these measures play a crucial role. The variations in approaches are influenced by the specific intervention types and the unique conditions and implementation at each site (Elton & Gruber, 1977).

When examining the connections between different actions and tools within the flood-risk system, it is important to consider the development of strategic options. Different approaches to interventions are discussed that involve combining options from different policy areas. Considering natural and societal trends can offer a more comprehensive perspective. Examples of natural trends include the continuous shift in climate patterns, while societal trends encompass changes in population dynamics and economic growth. Here, both are referred to as qualitative-quantitative scenarios. Such scenarios can be seen as envisioned futures over the flood risk system (Roselius, 1971).

4.3. FLOOD RISK MANAGEMENT PROCESS

Learning Objectives:

- Understand flood risk management optimization loop: analysis, assessment, reduction.
- Explore monitoring's role in flood risk management for adaptive strategies.
- Understand societal conditions influencing flood risk management decision-making.
- Grasp dimensions of flood risk management: content, context, process.

Further consideration is needed for the management of the risk of flood systems. Establishing connections between the analysis, assessment, and reduction of risks is essential for implementing a cohesive strategy in managing flood risks. Implementing these methods and tools into an operational system can be quite challenging due to their wide variations (Plate, 2002).

It is important to understand that the implementation of tasks and components by members of the community in flood risk management does not always adhere strictly to formal written guidelines. Instead, decisions and execution are influenced by various aspects of social behavior. Recognizing flood risk management requires a special focus on the practicality and actual application of scientific contributions in risk analysis, assessment, and reduction. The usability and relevance of these contributions in social discussions and decision-making processes are crucial (Albano et al., 2017).

Figure 4.4 presents a fundamental framework for managing flood risks, considering the connections and alignment with the decision-making and development process of society. The latter is positioned at the top as the primary element. Decisions and actions are made by the individuals involved in managing flood risks (Hsu et al., 2011). All the tasks and components mentioned earlier are focused on the process of making decisions and promoting development in society. Decision makers are faced with the task of addressing risk analysis, assessment, and reduction. To accomplish this, information provided by scientists and experts is required, or receive support from them. Decisions and actions are taken based on this information (Mai et al., 2020).

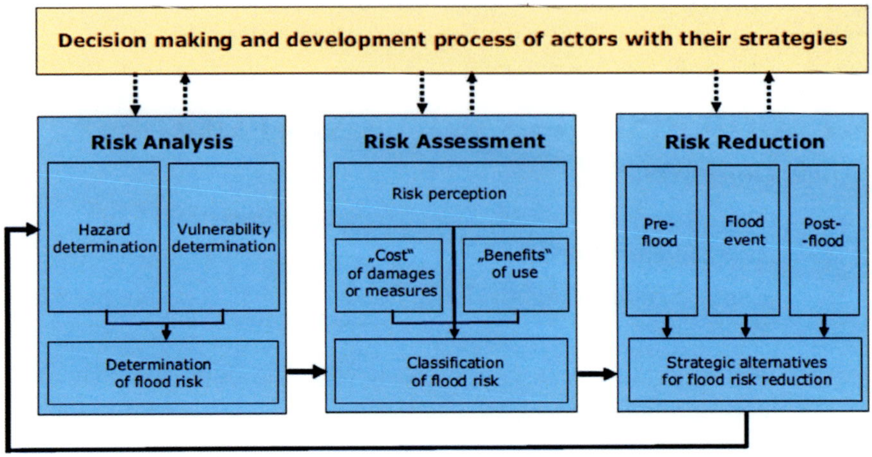

Figure 4.4. *Illustration of the basic framework of flood risk management.*

Source: Jiri Marsalek; Creative Commons License.

The framework is presented as a logical sequence. There is no consideration of time in this. Exploring further specifications is beyond the scope of this chapter. The subsequent paragraphs delve into the textual features and the process of social choice and development with greater depth (Apel et al., 2004).

4.3.1. Optimization and Monitoring of Flood Risks

Tasks and components of flood risk management have already been discussed earlier. Now, the issue at hand is how they both contribute to the broader aspect of managing flood risks. Each task and components are interconnected within the optimization loop. Optimization involves guiding the risk of flooding system towards a state of risk that aligns with acceptable levels of risk and requires minimal effort (Freeman, 1989). Using risk analysis, the flood risk system can be simulated. After conducting a thorough assessment, it becomes clear that certain areas need to be addressed to enhance the flood risk system. Developing options and alternatives to address risks that cannot be tolerable is a crucial aspect of risk reduction. These choices and alternatives are thoroughly examined about current trends and then simulated using risk analysis methods (Lu et al., 2020).

To effectively carry out risk analysis, assessment, and reduction, it is crucial to adopt a systematic approach. Effective simulations of the flood risk system require a collaborative approach to parameterization of model systems, assessment tools, and measures and instruments. It is essential to consider not only the intended outcomes but also the potential unintended consequences of specific decisions. Given the various impacts of flood management acts, the use of multi-criteria tools for risk evaluation is essential (Lund, 2002).

Did You Know?

Various cultures across the globe have long-held beliefs and stories surrounding floods. Throughout history, various stories have captured humanity's enduring fascination with the immense power of water and its ability to bring both devastation and renewal.

Scenarios and their analysis before the fact, particularly for the long term, often involve a significant level of uncertainty. Estimating the attainment of desired outcomes and the prevention of unintended consequences can only be partially achieved. Monitoring the actual progress of the risk of flood systems is crucial (Tasseff et al., 2019). Continuous gathering of information on the SPRC factors is necessary. Additionally, it is necessary to have information regarding the long-term maintenance of physical measures. It is important to also conduct evaluations of the effectiveness of measures and instruments used in past events. By closely monitoring the dynamics of the actual system and evaluating the effectiveness of interventions, valuable insights can be gained to enhance the accuracy of estimated impacts and make necessary adjustments to interventions (Yang et al., 2022).

4.3.2. Decision-Making and Development Process

As a foundation for the framework, it is assumed that the individuals involved in the development and decision-making procedure for flood risk management can serve as a 'managing entity.' It is worth noting that numerous decisions, particularly within catchments and possibly in coastal cells, are taken by various actors with different perspectives (Eisenhardt & Zbaracki, 1992). There may be separate development and decision-making processes that are not fully connected. In certain sectors, there are administrative units or actors at administrative levels who are taking on the role of managing entities. They are actively involved in analyzing, assessing, and reducing flood risks (Byrnes, 2002).

Given the critical nature of decision-making and development processes, it is necessary to gain a deeper understanding of the societal conditions surrounding them. It is crucial to consider the necessary elements for successful risk management, such as the need to integrate various stakeholders and choices into a comprehensive approach that encompasses the entire catchment or coastal area. Understanding of these matters remains restricted (Reymen et al., 2017). There is already a wealth of knowledge and understanding that can be gained from research in strategy and spatial planning. Hutter outlines three key aspects of creating strategies in flood risk management: the content, the context, and the methodological dimensions. The author highlights that the information about the tasks and components of flood risk management is just one aspect. To gain insight into how decisions are made by society, it is important to take into account the circumstances of the decision-makers and the steps involved in developing and executing strategies (Dos Santos et al., 2019).

The context dimension represents the conditions that actors in flood risk management consider as predetermined in a particular situation. The factors involved include external elements such as political, economic, and legal aspects, as well as internal factors like cultural influences. The process dimension highlights the variability in flood risk management procedures. In academic research, a common approach is to adopt a linear process model that focuses on a systematic, well-organized, analytical, and step-by-step procedure. Strategies in real-world conditions often do not follow a linear path during formulation and implementation. Gaining a deeper understanding of these processes and exploring different options could potentially enhance the overall effectiveness of managing flood risks. Continuity, implementation, ex-post evaluation, and reformulation of strategies are crucial in understanding the real dynamic within the flood risk system and the uncertainties involved in predicting the effects of decisions (Simon et al., 2006).

Learning from previous decisions is of special importance in this regard. Without additional details in this part, it becomes clear that flood risk management should not be limited to just one aspect. It is influenced by the societal factors that shape decision-making. The process dimension appears to be crucial for the successful implementation of strategies. The information provided sets the stage for closely monitoring the evolution of the actual flood risk system and enables adjustments to be made to previous objectives and goals. It may be a necessary condition for the co-evolutionary development between societal decisions and the ever-changing flood risk system. In addition, the context and process conditions are likely to play a crucial role in integrating fragmented decisions (Sternberg, 2003).

Practice Problem

In addressing flood risks in the city, a key challenge for company, Aqua Guard Solutions, is to develop a comprehensive flood risk management plan while simultaneously establishing an effective monitoring system to track changes in flood risks over time.

Solutions to Practice Problem

To tackle this challenge, Aqua Guard Solutions will adopt a multifaceted approach. Firstly, we will develop a flood risk management plan that integrates both structural and non-structural measures. This will include constructing flood barriers, enhancing drainage systems, and implementing early warning systems and community education programs. Additionally, we will prioritize collaboration with local authorities, stakeholders, and residents to ensure the plan's effectiveness and sustainability. Simultaneously, we will design and implement an advanced monitoring system leveraging cutting-edge technologies such as sensors, gauges, and remote sensing. This system will enable real-time data collection on rainfall patterns, river levels, soil moisture, and infrastructure conditions. By analyzing this data, we can identify vulnerabilities, track trends, and make informed decisions to mitigate flood risks effectively.

4.4. CHALLENGES FOR RESEARCH AND DEVELOPMENT

Learning Objectives:

- Understand dimensions of integration in flood risk management.
- Recognize challenges posed by temporal continuity in flood risk management.

The comprehensive and ongoing approach to flood risk management encompasses various pertinent factors, including physical processes and societal decision-making. In addition to acquiring more knowledge about individual things, the process of integration emerges as a key difficulty. Therefore, in the literature on flood risk management, the term "integrated" is occasionally employed as an adjective. Depending on the situation, there may be a combination of physical measures and other strategies involved, or a focus on different stages of flood management, such as before, during, and after a flood, to minimize risks (Saadatian et al., 2012). Originally, the concept of 'integration' referred to either (i) restoring something as a whole; or (ii) being involved in a larger entity. When it comes to flood risk management, integration is crucial. It involves considering all relevant aspects to gain a comprehensive understanding of how to effectively manage flood risks. Given the multitude of factors involved, it is important to consider various dimensions. There are two aspects to consider when studying flood risk assessment and management. The first aspect focuses on the physical processes involved in the flood risk system and how it gets spread out in space. The second aspect looks at the different stages of a flood, including the period before the flood, the actual flood event, and the aftermath. Additional aspects can be observed in the various approaches, as well as in different sectors, levels, and neighboring regions of decision-making (Rüegg et al., 2007).

Considering the definition of flood risk management presented here, all of these dimensions are inherently significant. Therefore, it appears that there is no longer a need for an explicit utilization of the term 'integrated.' However, understanding the theoretical, methodological, and technological aspects of integration is crucial for both research and practical applications in flood risk management. The focus is on developing comprehensive model systems to simulate the flood risk system. For working on incorporating these models into decision support systems and quantifying uncertainties via various approaches (Dadam & Reichert, 2009). Additionally, the results of the research show whether these results can be used in social decision-making processes or not. There is a growing recognition of the importance of these approaches in both research and practice. In addition to efforts to improve understanding of both natural and societal processes, advancements in integration can be built upon existing knowledge and further developed simultaneously (Smith et al., 1996).

Another significant hurdle in studying and implementing flood risk management is the ongoing nature of addressing the flood risk system. One aspect to consider is how these systems may change over time due to natural or societal factors, as well as intentional or unintentional interventions. Long-term predictions or scenarios involving changes are inherently associated with uncertainties. It is important to anticipate and quantify uncertainties in this context. Ex-post analyses can play a significant role in this context (Isenberg et al., 2011).

It is important to consider the ever-changing nature of flood risk and the challenges it presents when developing and executing strategies. The development process and decision must be guided by procedures that prioritize flexibility and the ability to adapt goals and interventions. It is crucial to have the capacity to gain insights from past choices and to be ready for unexpected circumstances (Amit et al., 1993).

SUMMARY

- The chapter has shown the requirements and possibilities to understand and deal with flood risk management as a continuous and holistic issue. Main tasks and components have been indicated and briefly explained.
- A basic framework provides a mental map for the systematization of the main components and their relations. It provides a means of navigation and navigating the scientific and practical contributions that are required for flood risk management. Moreover, it makes it possible to indicate whether all relevant aspects of site-specific flood risk management are being considered.
- The introduced framework stresses the meaning of the societal decision-making and development process as the managing procedure. Dedication of all tasks and components to this process appeared to be crucial. Furthermore, it is evident that flood risk management is not just restricted to information and options on the flood risk system. Instead, it happens within a societal context as a process of formulation and implementing strategies. Hereby, global change and unavoidable uncertainties of knowledge require a co-evolutionary understanding of steering and learning.
- Multiple dimensions of flood risk management ask for further research efforts to describe and make operational the framework of integration. Accordingly, knowledge can be expected to be crucial in contributing to more comprehensive and effective flood risk management.

REVIEW QUESTIONS

1. What is the fundamental difference between flood risk analysis and flood risk assessment?
2. How do cultural factors influence risk perception in flood risk assessment?
3. Can you explain the concept of "integrated" flood risk management and its significance?
4. Describe the components involved in flood risk reduction and provide examples of each.
5. Why is monitoring essential in flood risk management, and what does it entail?

6. Discuss the challenges associated with decision-making in flood risk management, particularly regarding temporal continuity.
7. How do societal conditions affect the decision-making process in flood risk management?

MULTIPLE CHOICE QUESTIONS

1. Which type of innovative material is designed to autonomously repair cracks in its structure?

a. Recycled Aggregates

b. Self-Healing Concrete

c. Shape Memory Alloys (SMAs)

d. Bamboo-reinforced Concrete

2. What does "flood risk" refer to in the context of flood risk assessment?

a. The likelihood of a flood occurring

b. The potential adverse effects of flooding on people and property

c. The duration of a flood event

d. The geographic extent of floodplains

3. What is the primary objective of flood risk management?

a. To prevent all floods from occurring

b. To eliminate all risks associated with flooding

c. To reduce the adverse impacts of flooding on human life and property

d. To increase the frequency of flooding events

4. What does flood risk analysis primarily involve?

a. Predicting future flood events

b. Assessing the perception of flood risks

c. Determining past, current, and future flood risks

d. Implementing flood protection measures

5. Which factor influences risk perception in flood risk assessment?

a. Climate change

b. Cultural background

c. Economic growth

d. Geographical location

6. What are examples of non-structural measures for flood risk reduction?

a. Construction of dams and levees

b. Implementation of early warning systems
c. Land-use planning and zoning regulations
d. Evacuation during flood events

Answers to Multiple Questions

1. (b); 2. (c); 3. (c); 4. (b); 5. (c); 6. (b)

REFERENCES

1. Akintoye, A. S., & MacLeod, M. J. (1997). Risk analysis and management in construction. *International Journal of Project Management, 15*(1), 31–38. https://doi.org/10.1016/S0263-7863(96)00035-X.
2. Albano, R., Mancusi, L., & Abbate, A. (2017). Improving flood risk analysis for effectively supporting the implementation of flood risk management plans The case study of "Serio" Valley. *Environmental Science & Policy, 75*, 158–172. https://doi.org/10.1016/j.envsci.2017.06.006.
3. Alexander, D. E. (2013). Resilience and disaster risk reduction: An etymological journey. *Natural Hazards and Earth System Sciences, 13*(11), 2707–2716. https://doi.org/10.5194/nhess-13-2707-2013.
4. Amit, R., Glosten, L., & Muller, E. (1993). Challenges to theory development in entrepreneurship research. *Journal of Management Studies, 30*(5), 815–834. https://doi.org/10.1111/j.1467-6486.1993.tb00327.x.
5. Apel, H., Thieken, A. H., Merz, B., & Blöschl, G. (2004). Flood risk assessment and associated uncertainty. *Natural Hazards and Earth System Sciences, 4*(2), 295–308. https://doi.org/10.5194/nhess-4-295-2004.
6. Baan, P. J., & Klijn, F. (2004). Flood risk perception and implications for flood risk management in the Netherlands. *International Journal of River Basin Management, 2*(2), 113–122. https://doi.org/10.1080/15715124.2004.9635226.
7. Benson, C., Twigg, J., & Myers, M. (2001). NGO initiatives in risk reduction: An overview. *Disasters, 25*(3), 199–215. https://doi.org/10.1111/1467-7717.00171.
8. Bergsma, E. (2019). The development of flood risk management in the United States. *Environmental Science & Policy, 101*, 32–37. https://doi.org/10.1016/j.envsci.2019.07.008.
9. Birkholz, S., Muro, M., Jeffrey, P., & Smith, H. M. (2014). Rethinking the relationship between flood risk perception and flood management.

Science of the Total Environment, 478, 12–20. https://doi.org/10.1016/j.scitotenv.2014.01.061.

10. Buchecker, M., Salvini, G., Di Baldassarre, G., Semenzin, E., Maidl, E., & Marcomini, A. (2013). The role of risk perception in making flood risk management more effective. *Natural Hazards and Earth System Sciences, 13*(11), 3013–3030. https://doi.org/10.5194/nhess-13-3013-2013.
11. Burningham, K., Fielding, J., & Thrush, D. (2008). 'It'll never happen to me': Understanding public awareness of local flood risk. *Disasters, 32*(2), 216–238. https://doi.org/10.1111/j.1467-7717.2008.01035.x.
12. Byrnes, J. P. (2002). The development of decision-making. *Journal of Adolescent Health, 31*(6), 208–215. https://doi.org/10.1016/S1054-139X(02)00303-7.
13. Catania, J. A., Kegeles, S. M., & Coates, T. J. (1990). Towards an understanding of risk behavior: An AIDS risk reduction model (ARRM). *Health Education Quarterly, 17*(1), 53–72. https://doi.org/10.1177/109019819001700107.
14. Cho, J., & Lee, J. (2006). An integrated model of risk and risk-reducing strategies. *Journal of Business Research, 59*(1), 112–120. https://doi.org/10.1016/j.jbusres.2005.03.004.
15. Cornell, C. A. (1968). Engineering seismic risk analysis. *Bulletin of the Seismological Society of America, 58*(5), 1583–1606. https://doi.org/10.1785/BSSA0580051583.
16. Cvetkovic, V. M., & Martinović, J. (2020). Innovative solutions for flood risk management. *International Journal of Disaster Risk Management, 2*(2), 71–100. https://doi.org/10.18485/ijdrm.2020.2.2.4.
17. Dadam, P., & Reichert, M. (2009). The ADEPT project: A decade of research and development for robust and flexible process support: Challenges and achievements. *Computer Science–Research and Development, 23*, 81–97. https://doi.org/10.1007/s00450-009-0055-6.
18. De Bruijn, K. M. (2004). Resilience and flood risk management. *Water Policy, 6*(1), 53–66. https://doi.org/10.2166/wp.2004.0007.
19. Diaconu, D. C., Costache, R., & Popa, M. C. (2021). An overview of flood risk analysis methods. *Water, 13*(4), 474. https://doi.org/10.3390/w13040474.
20. Dos Santos, P. H., Neves, S. M., Sant'Anna, D. O., De Oliveira, C. H., & Carvalho, H. D. (2019). The analytic hierarchy process supporting decision making for sustainable development: An overview of applications. *Journal of Cleaner Production, 212*, 119–138. https://doi.org/10.1016/j.jclepro.2018.11.270.

21. Eisenhardt, K. M., & Zbaracki, M. J. (1992). Strategic decision making. *Strategic Management Journal, 13*(S2), 17–37. https://doi.org/10.1002/smj.4250130904.

22. Elton, E. J., & Gruber, M. J. (1977). Risk reduction and portfolio size: An analytical solution. *The Journal of Business, 50*(4), 415–437. https://doi.org/10.1086/295964.

23. Fishburn, P. C. (1977). Mean-risk analysis with risk associated with below-target returns. *The American Economic Review, 67*(2), 116–126. https://doi.org/10.3386/w0228.

24. Foudi, S., Osés-Eraso, N., & Tamayo, I. (2015). Integrated spatial flood risk assessment: The case of Zaragoza. *Land Use Policy, 42*, 278–292. https://doi.org/10.1016/j.landusepol.2014.08.002.

25. Freeman, D. (1989). Teacher training, development, and decision making: A model of teaching and related strategies for language teacher education. *TESOL Quarterly, 23*(1), 27–45. https://doi.org/10.2307/3587502.

26. Gersonius, B., Zevenbergen, C., & van Herk, S. (2008). Managing flood risk in the urban environment: linking spatial planning, risk assessment, communication and policy. *Adaptive and Integrated Water Management: Coping with Complexity and Uncertainty, 5*(3), 263–275. https://doi.org/10.1007/978-3-540-75941-6_14.

27. Gouldby, B., Sayers, P., Mulet-Marti, J., Hassan, M. A. A. M., & Benwell, D. (2008, June). A methodology for regional-scale flood risk assessment. In *Proceedings of the Institution of Civil Engineers–Water Management* (Vol. 161, No. 3, pp. 169–182). Thomas Telford Ltd. https://doi.org/10.1680/wama.2008.161.3.169.

28. Hall, J. W., Meadowcroft, I. C., Sayers, P. B., & Bramley, M. E. (2003). Integrated flood risk management in England and Wales. *Natural Hazards Review, 4*(3), 126–135. https://doi.org/10.1061/(ASCE)1527-6988(2003)4:3(126).

29. Hartmann, T., & Driessen, P. (2017). The flood risk management plan: Towards spatial water governance. *Journal of Flood Risk Management, 10*(2), 145–154. https://doi.org/10.1111/jfr3.12077.

30. Hartmann, T., & Spit, T. (2016). Implementing the European flood risk management plan. *Journal of Environmental Planning and Management, 59*(2), 360–377. https://doi.org/10.1080/09640568.2015.1009629.

31. Hsu, W. K., Huang, P. C., Chang, C. C., Chen, C. W., Hung, D. M., & Chiang, W. L. (2011). An integrated flood risk assessment model for the property insurance industry in Taiwan. *Natural Hazards, 58*, 1295–1309. https://doi.org/10.1007/s11069-011-9736-5.

32. Isenberg, P., Elmqvist, N., Scholtz, J., Cernea, D., Ma, K. L., & Hagen, H. (2011). Collaborative visualization: Definition, challenges, and research agenda. *Information Visualization, 10*(4), 310–326. https://doi.org/10.1177/1473871611412817.
33. Klijn, F., Kreibich, H., De Moel, H., & Penning-Rowsell, E. (2015). Adaptive flood risk management planning based on a comprehensive flood risk conceptualization. *Mitigation and Adaptation Strategies for Global Change, 20*, 845–864. https://doi.org/10.1007/s11027-015-9638-z.
34. Kray, J., Li, K. Z., & Lindenberger, U. (2002). Age-related changes in task-switching components: The role of task uncertainty. *Brain and Cognition, 49*(3), 363-381. https://doi.org/10.1006/brcg.2001.1508.
35. Kriston, L., Scholl, I., Hölzel, L., Simon, D., Loh, A., & Härter, M. (2010). The 9-item shared decision-making questionnaire (SDM-Q-9). Development and psychometric properties in a primary care sample. *Patient Education and Counseling, 80*(1), 94–99. https://doi.org/10.1016/j.pec.2009.09.034.
36. Lechowska, E. (2022). Approaches in research on flood risk perception and their importance in flood risk management: A review. *Natural Hazards, 111*(3), 2343–2378. https://doi.org/10.1007/s11069-021-05022-0.
37. Li, K., Wu, S., Dai, E., & Xu, Z. (2012). Flood loss analysis and quantitative risk assessment in China. *Natural Hazards, 63*, 737–760. https://doi.org/10.1007/s11069-012-0184-2.
38. Lu, M., Liao, X., Yue, H., Huang, Y., Ye, H., Xu, C., & Huang, S. (2020). Optimizing the distribution of drone ports for emergency monitoring of flood disasters in China. *Journal of Flood Risk Management, 13*(1), e12593. https://doi.org/10.1111/jfr3.12593.
39. Lund, J. R. (2002). Floodplain planning with risk-based optimization. *Journal of Water Resources Planning and Management, 128*(3), 202–207. https://doi.org/10.1061/(ASCE)0733-9496(2002)128:3(202).
40. Lyu, H. M., Shen, S. L., Zhou, A., & Yang, J. (2019). Perspectives for flood risk assessment and management for mega-city metro system. *Tunnelling and Underground Space Technology, 84*, 31–44. https://doi.org/10.1016/j.tust.2018.10.018.
41. Mackenzie, F. D., Hirst, L. W., Battistutta, D., & Green, A. (1992). Risk analysis in the development of pterygia. *Ophthalmology, 99*(7), 1056–1061. https://doi.org/10.1016/S0161-6420(92)31737-7.
42. Mai, T., Mushtaq, S., Reardon-Smith, K., Webb, P., Stone, R., Kath, J., & An-Vo, D. A. (2020). Defining flood risk management strategies: A systems approach. *International Journal of Disaster Risk Reduction, 47*,

101550. https://doi.org/10.1016/j.ijdrr.2020.101550.

43. Marchi, L., Borga, M., Preciso, E., & Gaume, E. (2010). Characterization of selected extreme flash floods in Europe and implications for flood risk management. *Journal of Hydrology, 394*(1–2), 118–133. https://doi.org/10.1016/j.jhydrol.2010.07.017.
44. Merz, B., Aerts, J. C. J. H., Arnbjerg-Nielsen, K., Baldi, M., Becker, A., Bichet, A., & Nied, M. (2014). Floods and climate: Emerging perspectives for flood risk assessment and management. *Natural Hazards and Earth System Sciences, 14*(7), 1921–1942. https://doi.org/10.5194/nhess-14-1921-2014.
45. Merz, B., Hall, J., Disse, M., & Schumann, A. (2010). Fluvial flood risk management in a changing world. *Natural Hazards and Earth System Sciences, 10*(3), 509–527. https://doi.org/10.5194/nhess-10-509-2010.
46. Mitchell, V. W., & McGoldrick, P. J. (1996). Consumer's risk-reduction strategies: A review and synthesis. *International Review of Retail, Distribution and Consumer Research, 6*(1), 1–25. https://doi.org/10.1080/09593969600000001.
47. Morrison, A., Westbrook, C. J., & Noble, B. F. (2018). A review of the flood risk management governance and resilience literature. *Journal of Flood Risk Management, 11*(3), 291–304. https://doi.org/10.1111/jfr3.12315.
48. Mulholland, B., & Christian, J. (1999). Risk assessment in construction schedules. *Journal of Construction Engineering and Management, 125*(1), 8–15. https://doi.org/10.1061/(ASCE)0733-9364(1999)125:1(8).
49. Paté-Cornell, M. E. (1996). Uncertainties in risk analysis: Six levels of treatment. *Reliability Engineering & System Safety, 54*(2–3), 95–111. https://doi.org/10.1016/S0951-8320(96)00067-1.
50. Plate, E. J. (2002). Flood risk and flood management. *Journal of Hydrology, 267*(1–2), 2–11. https://doi.org/10.1016/S0022-1694(02)00135-X.
51. Prieto, C., Patel, D., & Han, D. (2020). Preface: Advances in flood risk assessment and management. *Natural Hazards and Earth System Sciences, 20*(4), 1045–1048. https://doi.org/10.5194/nhess-20-1045-2020.
52. Reymen, I., Berends, H., Oudehand, R., & Stultiëns, R. (2017). Decision making for business model development: A process study of effectuation and causation in new technology-based ventures. *R&D Management, 47*(4), 595–606. https://doi.org/10.1111/radm.12249.
53. Roselius, T. (1971). Consumer rankings of risk reduction methods. *Journal of Marketing, 35*(1), 56–61. https://doi.org/10.2307/1250565.
54. Rüegg, W. T., Quadranti, M., & Zoschke, A. (2007). Herbicide research

and development: Challenges and opportunities. *Weed Research, 47*(4), 271–275. https://doi.org/10.1111/j.1365-3180.2007.00572.x.

55. Rugman, A. M. (1976). Risk reduction by international diversification. *Journal of International Business Studies, 7*, 75–80. https://doi.org/10.1057/palgrave.jibs.8490689.
56. Saadatian, O., Sopian, K., Lim, C. H., Asim, N., & Sulaiman, M. Y. (2012). Trombe walls: A review of opportunities and challenges in research and development. *Renewable and Sustainable Energy Reviews, 16*(8), 6340–6351. https://doi.org/10.1016/j.rser.2012.07.012.
57. Schröter, K., Barendrecht, M., Bertola, M., Ciullo, A., da Costa, R. T., Cumiskey, L., & Sosa, J. (2021). Large-scale flood risk assessment and management: Prospects of a systems approach. *Water Security, 14*, 100–109. https://doi.org/10.1016/j.wasec.2021.100109.
58. Shah, M. A. R., Rahman, A., & Chowdhury, S. H. (2018). Challenges for achieving sustainable flood risk management. *Journal of Flood Risk Management, 11*(S1), S352–S358. https://doi.org/10.1111/jfr3.12230.
59. Short, J. F. (1984). The social fabric at risk: Toward the social transformation of risk analysis. *American Sociological Review, 49*(6), 711–725. https://doi.org/10.2307/2095526.
60. Simon, D., Schorr, G., Wirtz, M., Vodermaier, A., Caspari, C., Neuner, B., & Härter, M. (2006). Development and first validation of the shared decision-making questionnaire (SDM-Q). *Patient Education and Counseling, 63*(3), 319–327. https://doi.org/10.1016/j.pec.2006.04.012.
61. Skeem, J. L., & Monahan, J. (2011). Current directions in violence risk assessment. *Current Directions in Psychological Science, 20*(1), 38–42. https://doi.org/10.1177/0963721410397271.
62. Smith, M. A., McCaffrey, R. P., & Karp, J. E. (1996). The secondary leukemias: Challenges and research directions. *JNCI: Journal of the National Cancer Institute, 88*(7), 407–418. https://doi.org/10.1093/jnci/88.7.407.
63. Solin, L., & Skubincan, P. (2013). Flood risk assessment and management: Review of concepts, definitions, and methods. *Geographic Polonica, 65*(1), 23–44. https://doi.org/10.7163/GPol.2013.15.
64. Stapleton, J. M., & Halgren, E. (1987). Endogenous potentials evoked in simple cognitive tasks: Depth components and task correlates. *Electroencephalography and Clinical Neurophysiology, 67*(1), 44–52. https://doi.org/10.1016/0013-4694(87)90042-5.
65. Sternberg, R. J. (2003). The development of creativity as a decision-making

process. In R. K. Sawyer, V. John-Steiner, S. Moran, R. J. Sternberg, D. H. Feldman, J. Nakamura, & M. Csikszentmihalyi (Eds.), *Creativity and Development* (pp. 91–138). Oxford University Press. https://doi.org/10.1093/acprof:oso/9780195147300.003.0006.

66. Tasseff, B., Bent, R., & Van Hentenryck, P. (2019). Optimization of structural flood mitigation strategies. *Water Resources Research, 55*(2), 1490–1509. https://doi.org/10.1029/2018WR024228.

67. Thaler, T., & Hartmann, T. (2016). Justice and flood risk management: Reflecting on different approaches to distribute and allocate flood risk management in Europe. *Natural Hazards, 83*(1), 129–147. https://doi.org/10.1007/s11069-016-2305-1.

68. Thumerer, T., Jones, A. P., & Brown, D. (2000). A GIS-based coastal management system for climate change associated flood risk assessment on the east coast of England. *International Journal of Geographical Information Science, 14*(3), 265–281. https://doi.org/10.1080/136588100240830.

69. Tran, P., Marincioni, F., Shaw, R., Sarti, M., & Van An, L. (2008). Flood risk management in Central Viet Nam: Challenges and potentials. *Natural Hazards, 46*(1), 119–138. https://doi.org/10.1007/s11069-007-9200-4.

70. Tsakiris, G. J. (2014). Flood risk assessment: Concepts, modeling, applications. *Natural Hazards and Earth System Sciences, 14*(5), 1361–1369. https://doi.org/10.5194/nhess-14-1361-2014.

71. Vorogushyn, S., Bates, P. D., de Bruijn, K., Castellarin, A., Kreibich, H., Priest, S., & Merz, B. (2018). Evolutionary leap in large-scale flood risk assessment needed. *Wiley Interdisciplinary Reviews: Water, 5*(2), e1266. https://doi.org/10.1002/wat2.1266.

72. Waghwala, R. K., & Agnihotri, P. G. (2019). Flood risk assessment and resilience strategies for flood risk management: A case study of Surat City. *International Journal of Disaster Risk Reduction, 40*, 101155. https://doi.org/10.1016/j.ijdrr.2019.101155.

73. Wald, M. S., & Woolverton, M. (1990). Risk assessment: The emperor's new clothes? *Child Welfare, 69*(2), 483–511.

74. Weichselgartner, J., & Pigeon, P. (2015). The role of knowledge in disaster risk reduction. *International Journal of Disaster Risk Science, 6*(2), 107–116. https://doi.org/10.1007/s13753-015-0052-7.

75. Yang, F., Xiong, S., Ou, J., Zhao, Z., & Lei, T. (2022). Human settlement resilience zoning and optimizing strategies for river-network cities under flood risk management objectives: Taking Yueyang City as an example. *Sustainability, 14*(15), 9595. https://doi.org/10.3390/su14159595.

76. Zeleňáková, M. (2011). Flood risk assessment and management in Slovakia. *WIT Transactions on Ecology and the Environment, 146*, 61–70. https://doi.org/10.2495/RM110061.

CHAPTER 5

FLOOD CONTROL MEASURES AND MANAGEMENT

LEARNING OBJECTIVES

At the end of this chapter, readers will be able to:

1. Understand floodplain and its significance in flood management.
2. Understand the hydrologic and hydraulic analysis of floods.
3. Analyze the role of floodways and floodway fringes in floodplain management.
4. Evaluate structural methods, such as reservoirs, for flood damage reduction.
5. Explore the environmental impacts of channel modifications for flood control.
6. Know the nonstructural methods, like voluntary buyouts, for flood damage reduction.
7. Comprehend and categorize different types of flood damage.
8. Understand methodologies for estimating flood damage and risk-based design.

INTRODUCTORY EXAMPLE

Riverfront Engineering Solutions, Inc. is a leading engineering firm specializing in flood control measures and management. With expertise in hydrology, hydraulics, and urban planning, the company assists communities in developing comprehensive strategies to mitigate flood risks and minimize damages.

The city of Rivertown, situated along the banks of a major river, has experienced recurrent flooding events during heavy rainfall seasons. These floods have resulted in significant property damage, infrastructure disruptions, and displacement of residents. In response to these challenges, Rivertown has enlisted the expertise of Riverfront Engineering Solutions, Inc. to develop a comprehensive flood control plan.

Riverfront Engineering Solutions, Inc. begins by conducting a thorough hydrological analysis to assess the frequency and magnitude of floods in the region. Using advanced modeling techniques and historical data, they identify the 100-year floodplain and evaluate the potential impacts of various flood scenarios on the city's infrastructure and population.

Based on their analysis, Riverfront Engineering Solutions, Inc. proposes a multi-faceted approach to flood control, incorporating both structural and nonstructural measures. This includes the construction of levees, floodwalls, and retention basins to protect key areas from inundation, as well as the implementation of zoning regulations and land use planning strategies to limit development in high-risk flood zones.

Through collaboration with local authorities, community stakeholders, and environmental agencies, Riverfront Engineering Solutions, Inc. develops a comprehensive flood control plan tailored to the unique needs and challenges of Rivertown.

UNIT INTRODUCTION

Floods are a common occurrence in the natural history of our planet. Flooding can happen in various locations, including:

- Ground-failure areas;
- Coastal areas;
- Alluvial fans;
- Lakes, rivers, and streams;
- Areas with structural measures;
- Areas with surface runoff and inadequate drainage.

Human Settlements and activities have historically gravitated towards floodplains. Their implementation has often disrupted the natural flow of floodplain processes, resulting in significant disruptions and hardships for human populations (Li et al., 2016). This chapter offers a thorough examination of floodplain management strategies, covering the understanding of floodplains, analysis of hydrology and hydraulics, and the identification of floodways and fringes. The text delves deeper into the regulatory framework management policies that are essential for ensuring effective floodplain management. In addition, the course explores the incorporation of stormwater management techniques to improve flood resistance in both urban and rural areas. Exploring floodplain management, this unit explores a range of techniques for flood control, such as the use of flood control reservoirs, modifications to channels, and voluntary buyouts of characteristics at risk of flooding. The unit concludes by examining flood damage estimation methods, which are crucial for minimizing economic losses and optimizing flood control measures caused by flooding events. This includes studying risk-based analyses, damage relationships, and expected damages (Miguez & de Magalhães, 2010).

5.1. FLOODPLAIN MANAGEMENT

Learning Objectives

- Understand the definition and characteristics of floodplains in various settings.
- Know the methodologies and significance of hydrologic and hydraulic analyses in flood management.
- Understand the concept of floodways and floodway fringes and their role in floodplain management.
- Gain insights into stormwater management strategies and their integration with floodplain management.

5.1.1. Floodplain Definition

A floodplain is an area of land that is typically dry but becomes flooded during periods of heavy rainfall or when lakes, bays, rivers, streams, or oceans overflow their banks. Flooding is often caused by the overflow of streams and rivers, as well as abnormally high tides during severe storms. The floodplain encompasses both the entire width of narrowed ravines and the expansive areas across streams in wide, flat valleys. As depicted in Diagram 5.1, a channel and floodplain are essential components of the natural flow of a stream (Smemoe et al., 2007). Flow over the river's capacity is carried by the floodplain, and the degree of flow over the floodplain increases with the discharge. Floodplains can be defined as a regulatory perspective or a natural geological feature. For management and safety reasons, the standard used format in the US is the 100-year floodplain. Flooding is a concern that extends beyond riverine and coastal areas. There is also a need to address the issue of floods that are caused by factors such as ice jams, alluvial fans, unstable channels, mudflows, and subsidence (Krizek & Sobr, 2006). Alluvial fans are known for their distinctive shape, formed by the accumulation of various materials such as sediments, boulders, and gravel. These materials are typically eroded from slopes of mountains and then transported by natural processes like flood flows, debris flows, movement, sediment erosion, and channel migration. This process is illustrated in Figure 5.1. Alluvial fans can be found in various regions across the globe. Flood control measures and management are widely implemented in various states across the United States, including Arizona, Idaho, Montana, Utah, New Mexico, Nevada, Washington, California, and Wyoming (Croke et al., 2016).

5.1.2. Hydrologic and Hydraulic Analysis of Floods

Analysis of floods is essential for the effective planning, management, and design, of various facilities, such as hydro systems located in floodplains or watersheds.

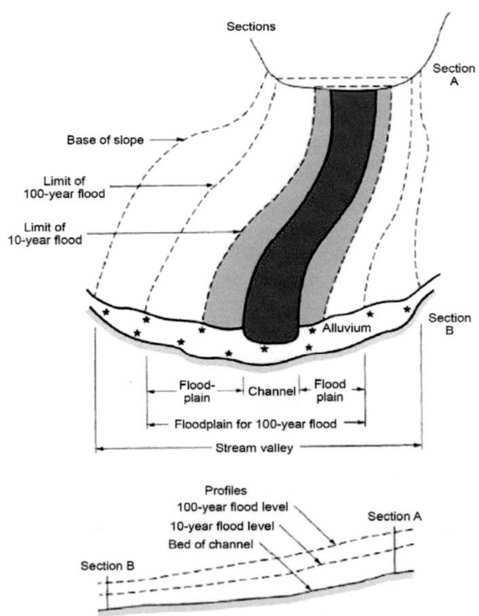

Figure 5.1. *Illustration of typical sections and profiles in an unobstructed reach of stream valley.*

Source: Waananen; Creative Commons License.

These evaluations are necessary to determine potential areas of inundation, sizing of channels, flood depths and heights, levee heights, right-of-way limits, highway crossings, culvert design, and various other factors (Anselmo et al., 1996).

Commonly needed criteria involve:

1. **Research on Floodplain Information:** Documenting details about various flood events, including those with different frequencies like the 10-, 100-, and 500-year events.
2. **Assessment of Potential Land-Use Options for the Future:** Examining various flood events with various frequencies, the potential for flooding is assessed, as the resulting damage, and the impact on the environment for current and future land uses.
3. **Assessment of Measures Aimed at Reducing Flood-Related Losses:** Examining various flood events with various frequencies to assess the extent of flood damage reduction linked to specific design flows.
4. **Studies on Design:** Examining flood incidents to determine the appropriate size of facilities to ensure their reliability and prevent any potential failures.
5. **Operational Studies:** Assessing the capability of a system to handle the requirements imposed by flood events (Gül et al., 2010).

The techniques employed in hydrologic and hydraulic analysis are influenced by the objectives and extent of the project, as well as the accessibility of data. Here is a schematic of the components involved in conducting a hydraulic and hydrologic analysis for floodplain studies, as shown in Figure 5.2. When it comes to analyzing floodplains, there are two main methods: rainfall-runoff analysis and flood-flow frequency analysis.

When there is enough data on historical yearly instantaneous peak discharges, it becomes possible to conduct flood-flow frequency analysis. This analysis helps determine the peak discharges for different return periods. Alternatively, it is necessary to conduct a rainfall-runoff analysis by utilizing either a historical storm or a design storm specific to a given return period to create a storm-runoff hydrograph (Aksoy et al., 2016) (Figure 5.3).

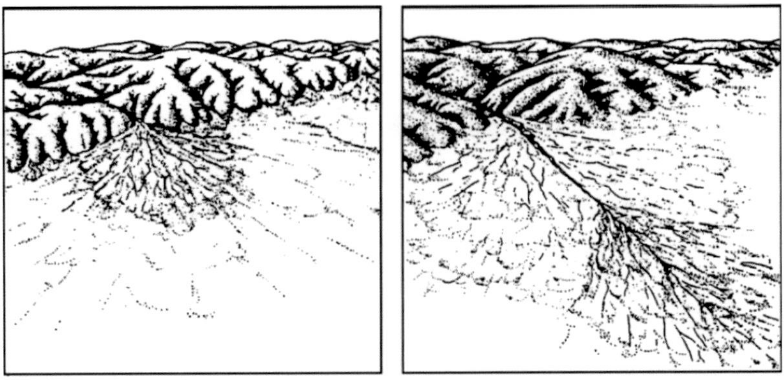

Figure 5.2. *Illustration of two types of alluvial fans.*

Source: W. B. Bull; Creative Commons License.

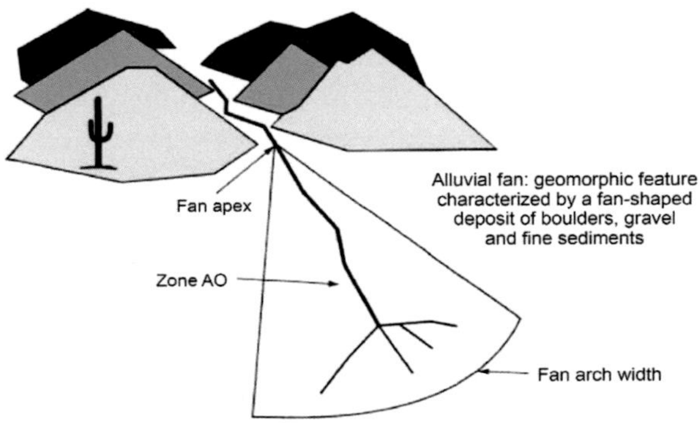

Figure 5.3. *Illustration of an alluvial fan system.*

Source: Larry Ways; Creative Commons License.

Water-surface heights can be determined through a steady-state water-surface profile analysis when only peak discharges are available. Alternatively, one can choose the peak discharges from storm-runoff hydrographs that have been generated. To provide a more thorough analysis, an unsteady-flow analysis can be conducted using a hydraulic routing model (Aureli et al., 2021). This analysis requires the storm-runoff hydrograph and allows for a more precise determination of maximum water-surface heights. The unsteady-flow analysis also offers a wealth of detailed information, including the hydrographs of routed discharge at different points along a river reach (Vangelis et al., 2022) (Figure 5.4).

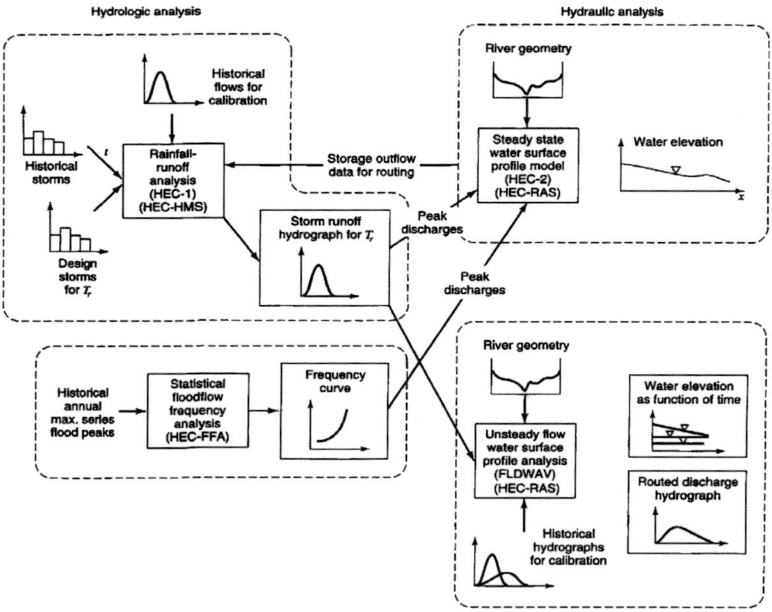

Figure 5.4. *Representation of the components of a hydrologic-hydraulic floodplain analysis.*

Source: W. B. Bull; Creative Commons License.

5.1.3. Floodways and Floodway Fringes

Building on floodplains, like through the use of man-made fill material, diminishes the ability of streams to carry floodwaters, raises the water levels during floods, and amplifies the risks of flooding in surrounding areas. One aspect of managing floodplains involves finding a balance between the economic benefits of developing the floodplain and the potential increase in flood risk that comes with it. Figure 5.5 illustrates the division of the 100-year flood area into a floodway and a floodway fringe, as per (FEMA) Federal Emergency Management Agency studies (Hosseinzadeh et al., 2022). The floodway refers to the stream channel and the surrounding floodplain areas that need to be protected from any development or encroachment. This is crucial to ensure that the 100-year flood can flow through without causing significant increases in flood heights (Walesh, 1979).

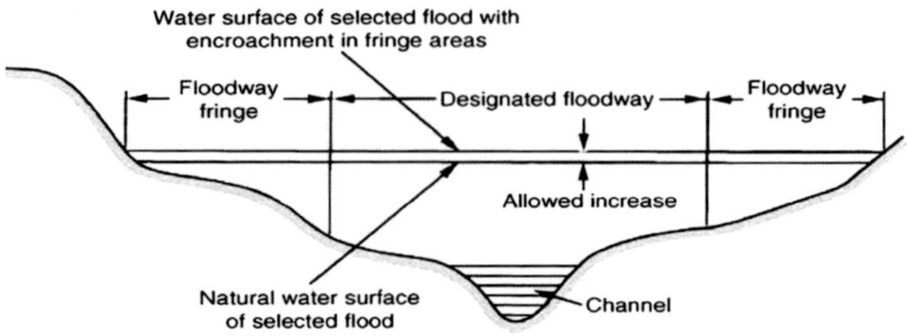

Figure 5.5. *Illustration definition of floodway and floodway fringe.*

Source: Daniel P. Lock; Creative Commons License.

The floodway fringe refers to the space between the designated floodway limit and the limit of the selected flood. The floodway limit is established to ensure that any development within the floodway fringe does not cause a significant rise in flood levels. In the United States, it is common to refer to the 100-year flood, with a standard 1-ft allowable increase. A flood 1.0 ft high can be maintained without generating dangerous velocities. The floodway fringe refers to the area of the floodplain that can be blocked without causing the water level of the 100-year flood to rise by over one foot at any location (Rahmani et al., 2020).

Two different types of maps, floodplain inundation maps, specifically flood-prone area maps and flood-hazard maps, have been utilized. Maps of flood-prone areas indicate regions that are at risk of flooding due to their proximity to bodies of water such as rivers, streams, bays, or oceans. This information is easily accessible and helps in identifying potential flood zones. A study of flooding at a specific location can provide valuable information on the extent of inundation, as shown in Figure 5.6 for Clear Creek, which is near Houston, Texas. Flood-hazard maps are frequently utilized in floodplain data reports and necessitate periodic updates to reflect any alterations in the channels, floodplains, and upstream regions (Liebman, 1973).

5.1.4. Floodplain Management and Floodplain Regulations

Floodplain management, as defined by the National Flood Insurance Program (NFIP) by rules overseen by FEMA, refers to the implementation of a comprehensive set of measures aimed at reducing the impact of flooding. These measures include, but are not restricted to, emergency preparedness plans, flood control infrastructure, and regulations about the use and development of flood-prone areas. Floodplain management policies are the optimal approach to mitigate future flood damage in emerging communities that are already aware of flood threats (Hornung et al., 2019).

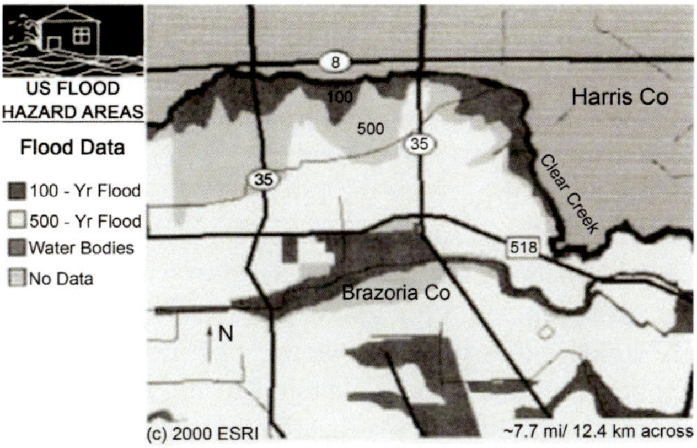

Figure 5.6. *Illustration of floodplain map for a portion of clear creek in Brazoria County near Houston, Texas.*

Source: Lawrence, K.; Creative Commons License.

Floodplain management examines issues that have emerged in developed regions and potential issues that can be anticipated because of upcoming changes. There are several fundamental approaches to managing floodplains:

- Actions to decrease vulnerability to floods;
- Actions that alter the flood dynamics;
- Actions that support communities and individuals managing flood events.

Floodplain regulation plays a crucial role in a comprehensive floodplain management program, especially in areas that are still developing and can regulate future development (Olsen, 2006).

Did You Know?

Certain animal species have developed distinct survival strategies to endure flooding events. As an illustration, the Amazon River dolphins, also referred to as the pink river dolphins, have the remarkable ability to navigate through flooded jungles throughout the rainy season if their habitat expands.

An important aspect of floodplain regulation involves determining the flood-hazard area, typically referred to as the 100-year floodplain, as well as the floodway. The floodway encompasses the stream channel and the surrounding land areas that need to be set aside to safely accommodate the designed flood, ensuring that the water surface does not rise beyond a specified limit. As an illustration, the National Flood Insurance Program (NFIP) sets a maximum allowable rise of one foot. However, in certain cases, a smaller increase may be more suitable (Burby, 2001). Implementing a comprehensive floodplain management strategy that prioritizes the removal of development from flood-prone areas can be a significant stride toward effective flood control. Preventing the placement of obstructions like fill and detention basins in the flood-hazard area is a positive advancement. A floodplain serves as a conveyance mechanism and temporarily holds excess water. Placing obstructions in the floodplain removes the temporarily stored areas, leading to higher hydraulic heads and increased flood levels regardless of the location of the floodplain developments (Kiedrzyńska et al., 2015).

5.1.5. Storm Water Management and Floodplain Management

Effective stormwater management strategies tend to be effective once carried out at the beginning of development within a region and must be administered within the framework of a comprehensive land-use planning process. Implementing a stormwater management strategy to address stream deterioration caused by uncontrolled growth is a difficult task. Stormwater detention programs can vary in their effectiveness. In certain situations, it may not have a significant impact because the flood peak resulting from detention decreases as the flood moves downstream. Additionally, the rise in total runoff caused by development can contribute to the overall mass of the flood wave (Braden & Johnston, 2004). The downstream impact of multiple detention basins would primarily be to slightly postpone the arrival of the flood peak by a few hours, with minimal or no impact on reducing the maximum flow rate. One approach to address this issue is to implement long-term retention strategies. Effective management of flood control measures requires careful and comprehensive planning before any development takes place to mitigate the impacts of increased peak flood discharges. Preserving undeveloped areas, especially those in the floodplain, through zoning can prove to be a highly effective measure (Haris et al., 2016).

Stormwater management along with floodplain management are typically treated as distinct programs, but both of them intersect in certain areas. For example, the construction of detention basins in floodplains presents unavoidable challenges. Detention basins are strategically implemented in various regions across the United States. It is advisable to locate detention basins outside of the floodplain, especially for smaller streams with smaller drainage areas. During such instances, the development site and the floodplain are both impacted by the same storms at the same time. But simply, the period when the detention basin is required to hold stormwater from development coincides with the time when the floodplain is already updated, and the detention basin's site would already be filled with floodwater. If the flood in the development aligns with other flooding in the floodplain, detention storage space in the floodplain proves to be ineffective. Another factor that has reduced the efficiency of detention basins within many regions is the significant amount of fill that comes with development. Implementing appropriate regional detention and retention systems, as well as enhancing drainage systems might be necessary to address drainage issues (Newcomer et al., 2014).

5.2. STRUCTURAL AND NONSTRUCTURAL METHODS OF FLOOD CONTROL

Learning Objectives:

- Understand the distinction between structural and nonstructural flood control methods.
- Understand the various structural methods including reservoirs, levees, and diversions.
- Know about nonstructural approaches like voluntary property buyouts and flood warning systems.

Flooding occurs when the combination of hydrological and topographical factors in floodplains leads to excessive water flow, causing the channel banks to overflow and resulting in the water spreading over the floodplain. During significant floods, the floodplain serves as both a means of transporting floodwater and a temporary storage area for the excess flow. The primary channel typically follows a defined path as it winds its way through the floodplain, transporting minimal amounts of water. Typically, the overbank flow is shallower and slower than the flow within the channel (James, 1965).

The goal of flood control is to minimize or mitigate the adverse effects of flooding. Alternative methods that modify the flow of floodwater are commonly known as flood-control facilities and include engineering structures or modifications. Structural measures for flood control, such as reservoirs, levees or dikes, diversions, and channel modifications, are constructed with careful consideration of the flood characteristics (Faisal et al., 1999). Nonstructural flood-control measures that alter the vulnerability of floodplains typically involve minor engineering works. Nonstructural measures aim to alter the damage potential of permanent facilities and minimize potential damage over a flood event. Other measures that can be implemented include flood-proofing, the establishment of flood warning systems, and the implementation of land-use controls. Implementing structural measures typically involves significant financial investment. Floodplain management considers a holistic approach to various measures, including engineering, nonstructural, and administrative, to minimize losses caused by flooding on a large scale (Meyer et al., 2012).

5.2.1. Structural Methods for Flood Damage Reduction

Throughout much of its long history, the US has consistently focused on implementing structures to manage and mitigate floods. The structural approach gained prominence in the 1930s following the enactment of the 1936 Flood Control Act and the establishment of the Federal Crop Insurance Corporation. The federal government took charge of the nation's flood prevention efforts and provided compensation to those affected by the floods. In 1940, the government took on the responsibility of funding and overseeing the construction and upkeep of dams, channel modifications, and rectification projects across all navigable waters in the United States (USEOP, 1994). Significant progress in fluid mechanics, infiltration, statistical hydrology, hydrologic systems, evaporation analysis, and flood routing during the 1940s greatly facilitated the design and construction of large-scale control structures, resulting in highly efficient flood control measures (Davar et al., 2001).

There are various options for managing floods through structural measures. These include modifying channels, using reservoirs or detention/retention ponds to control flood flows, constructing levees or floodwalls to contain floodwaters, and diverting flood flows to alternate floodways or detention ponds (Kreibich et al., 2015). Every method has its own set of conditions where it can be effective, along with its own set of pros and cons. The failure of levees in New Orleans through Hurricane Katrina serves as a stark reminder of the significance of selecting suitable structural methods to safeguard against various foreseeable circumstances. Here is a brief overview of these kinds of structural alternatives (Thampapillai & Musgrave, 1985).

5.2.1.1. Flood Control Reservoirs

Flood control reservoirs offer a highly efficient method for managing and regulating flood flows. Various types of basins can be implemented, ranging from small ponds constructed next to highways or road culverts to large reservoirs equipped with control structures capable of containing massive amounts of water. Regardless of its size, all reservoirs have the same purpose of providing extra storage to reduce the peak flow during a storm event (Cheng

& Chau, 2004). There are two main types of flood control reservoirs: detention and retention. Detention reservoirs or just ponds are specifically engineered to temporarily store excess runoff before gradually releasing it back into the natural watercourse. Retention ponds are created to store excess water for longer periods, typically for enhancing the landscape, supporting agriculture, meeting water demands, or serving other specific needs. Very large retention structures were also known as storage reservoirs. These storage structures are commonly linked to big reservoirs and are frequently constructed not only for protection from floods but also for water distribution purposes (Ren et al., 2017).

Did You Know?

Floods have the potential to uncover valuable discoveries. Following a flood at Venice, Italy, in 1966, a multitude of historical items and artifacts which had become lost or buried over time were revealed as the waters decreased.

The effectiveness of flood protection provided by a reservoir at a specific point on a stream depends mainly on the proportion of the watershed that lies upstream of the secured point, which is determined by the placement of the reservoir framework. The decrease in this reduction becomes more apparent as water from the surrounding watershed begins to flow into the reservoir and as the distance downstream from the reservoir grows. The distance can vary and is influenced by factors like the number and size of tributaries and the shape of the watershed. To be effective, it is important to have a reservoir site that is located at a suitable distance from the upstream area. One notable advantage of this approach is that it allows for the construction of a smaller reservoir upstream of flood-prone areas. This not only provides effective flood protection but also helps to minimize excavation costs and reduce environmental impact (Gotoh et al., 2011).

Practice Problem

In a region prone to seasonal flooding, where the historical peak flow is 10,000 cubic feet per second (cfs) occurring once every 20 years, local authorities are considering the construction of a flood control reservoir to alleviate flood impacts. The proposed reservoir is designed with a storage capacity of 50,000 acre-feet. Students are tasked with determining the design flood volume the reservoir needs to accommodate, calculating the frequency of occurrence for the design flood event, and evaluating the effectiveness of the proposed reservoir in mitigating flood risks.

Solutions to Practice Problem

To calculate the design flood volume, students utilize the formula for peak flow multiplied by recurrence interval, yielding a design flood volume of 200,000 cubic feet. This indicates that the expected flood event occurs every 20 years. Next, evaluating the reservoir's effectiveness involves comparing its storage capacity to the design flood volume. Converting the reservoir's storage capacity from acre-feet to cubic feet yields 2,178,000,000 cubic feet, indicating that it surpasses the design flood volume. Therefore, the reservoir demonstrates efficacy in mitigating the impact of the design flood event by providing sufficient storage capacity. Through this exercise, students gain practical insight into the functionality and assessment of flood control reservoirs in reducing flood damage.

On the other hand, the location of reservoirs can pose challenges when it comes to implementing extensive flood protection measures, especially when there are limited suitable sites available. This is because the construction of extremely large reservoirs becomes necessary to offer substantial downstream protection. Reservoirs offer minimal protection during severe storms in regions with gentle slopes and small natural channels. There are also some notable disadvantages to consider, such as the challenge of finding suitable sites to dispose of the excavated material and the issue of aesthetics whenever the constructed pond cannot be utilized for flood control reasons (Prékopa & Szántai, 1978).

5.2.1.2. Channel Modifications and Environmental Impacts

Stream channel modifications are implemented to enhance the flow features and capacity of a natural waterway. Modifying the hydraulic factors of channel slope, depth, width, and roughness can enhance the conveyance and holding capacity of a natural channel. The enhanced channel's improved hydraulic efficiency leads to higher flow velocities, resulting in lower flood phases or water surface heights during a storm event (Schoof, 1980).

These modifications, commonly known as "channelization," have been widely implemented across America for many years, yielding positive outcomes. This method is widely used to manage the flow of water in winding rivers found in the southern and southeast coastal plains. Nevertheless, the channeling approach has consistently sparked debate due to potential negative impacts on the environment that may outweigh any benefits. Channelization allows for the management of flooding in natural waterways, safeguarding established urban, agricultural, and industrial areas. Furthermore, channel projects have the potential to stimulate economic growth within the floodplain. Channelization projects are often implemented for various purposes, such as flood control, wetland drainage, navigation improvement, and prevention of bank erosion and channel migration. These projects also serve to safeguard neighboring property (Hupp et al., 2009).

Channelization has proven to be highly advantageous, offering benefits to various sectors of the economy. It has effectively controlled flooding in the implemented areas, benefiting transportation, urban, agriculture, and other industries. The effectiveness of flood control measures and management is evident in the modifications made to over 2,00,000 miles of this nation's waterways between 1820 and 1970.

Until the mid-1970s, channelization became the preferred alternative for several flood control agencies due to its track record of success. Nevertheless, increasing worries about the lasting efficiency of channeling and the ecological consequences caused by this method halted the widespread implementation of such projects. The channeling efforts of the Soil Conservation Service (SCS) and the COE became a topic of intense debate during the late 1960s in the United States. Congress hired Arthur D. Little, Inc., to conduct a thorough nationwide survey of the environmental impacts of these projects and present the results to the Council on the Quality of the Environment. This influential report would spark widespread discussion on the most effective strategies for managing and mitigating floods in the coming decades. It is undeniable that this report will have an immediate and far-reaching effect on projects aimed at channelizing waterways (Keller, 2020).

The report published in 1973 indicated that the channelization tasks it examined were conservatively estimated in terms of their direct benefits. The study found that most flood channel alterations were functioning as intended, resulting in a significant reduction in annual damage. The investment of approximately $15 billion in flood protection measures across the country has proven to be effective, saving an estimated $1 billion in damages each year. Certain projects, unfortunately, were not properly planned and did not deliver sufficient effectiveness in controlling floods for the intended design storm. The channel projects near Houston, including the Brays and White Oak Bayous, had been built to manage significant storm events. However, their capacities were significantly diminished due to the encouragement of floodplain development. Downstream communities have been significantly impacted by the effects of channelization, as seen in Houston and other major cities. With the progress of development, the speed at which flows are channeled tends to intensify the peak flows of upstream tributaries, thus worsening the conditions downstream.

Several channel modifications were poorly designed and executed, resulting in significant environmental consequences that outweighed the limited flood control advantages they offered (Elosegi & Sabater, 2013).

The Little report highlighted the importance of thoroughly evaluating the environmental impacts of channel work before implementing any channelization plan. Some measures can be implemented to address the issue:

- Draining of nearby wetlands;
- Cutting off oxbows and stream meanders;
- Lowering of groundwater levels, resulting in reduced stream recharge;
- Clearing of floodplain bottomland hardwoods and destruction of wildlife habitat;
- Increased erosion of stream banks, leading to stability problems;
- Increased downstream deposition of silt and sedimentation;
- Increased downstream flooding as a result of increased upstream velocities;
- Impacts on the stream's and receiving body's aquatic life;
- Reduced aesthetics and loss of visual amenity.

The conclusions of the little study in the 1970s redirected subsequent research towards the advancement of technology and science to accurately measure the environmental effects of channelization. Over the next 20 years, numerous scientific studies were published on the effects of channelization, encompassing both hydraulic and environmental aspects.

5.2.1.3. Diversion Channels, Levees, and By-Pass Channels

Diverting the flow of water is a common technique to redirect floodwaters away from areas that are susceptible to flooding. As a result, there is a decrease in the intensity of water flow in the specific location. The approach only offers protection downstream, with the level of protection decreasing like reservoirs. Diverting flows are commonly done alongside reservoirs located far from a main channel, as it is necessary to contain or redirect the diverted water around the affected area. One major limitation is the need for ample right-of-way to effectively channel diverted flows into adjacent basins (Surian et al., 2009).

Levees are constructed as linear structures, typically made of earthen embankments or reinforced concrete walls. And are built parallel to the mainstream to contain a designated overbank flow. These structures are typically situated near the natural floodway or on each side of the river, offering security to all of the natural floodplains. This approach is widely recognized as one of the earliest and most used methods of localized flood protection across the globe, including in the United States (Dogruoz et al., 2005). It serves as a crucial component in managing and mitigating the impact of floods. Levees are designed to enhance capacity and provide protection against flood damage in the surrounding river corridor if the water level remains below the highest point of the structure. One major drawback is that if they are overtopped, the resulting flood damage is as severe as though the structure was not there at all (Bouchakour et al., 2022). Considering the past patterns of development in flood-prone areas, particularly with the perceived protection offered by floodwalls, the resulting damage often surpasses what would have happened if there were no levees in place. There are some downsides to consider, such as higher flow rates downstream and the subsequent increase in flood damages. There is also the need for significant right-of-way space and potential environmental impacts, like what occurs with channelization. The unprecedented convergence of weather conditions resulted in the devastating flood of 1993, which caused extensive damage to the levees across the Mississippi River (Jolly et al., 2017).

The severe flooding of 2011 was the most recent in a series of floods that have affected the Mississippi basins. Initial projections suggested that the potential costs of the damage could amount to approximately 4 billion dollars. A significant amount of damage was inflicted upon croplands due to the activation of spillways, which was necessary to protect cities like Memphis, Tennessee, and New Orleans, Louisiana. Unlike the floods in 1993, the event in 2011 was compounded by melting thick winter snows along with excessive rainfall (Droz et al., 2020). Bypass channels are utilized to confine and transfer a part of storm flows away from or within a highly flood-prone or ecologically sensitive area. These channels have similar limitations to the enhancements made to the mainstream or floodway. The broad applicability of this method is hindered by important right-of-way and property acquisitions. In addition, there are often substantial expenses involved in excavating channels in areas without existing streambeds (Carver, 1969).

5.2.2. Nonstructural Methods for Flood Damage Reduction

Methods are employed to minimize the potential harm to a structure or facility located in a flood-prone area by modifying the utilization of land susceptible to flooding. This approach presents a clear contrast to its structural counterpart, as it focuses on minimizing the flood hazard through modifying flood flows. The effectiveness of nonstructural measures in reducing damage potential can vary greatly depending on the specific mechanism implemented (Hayes, 2004).

There are numerous nonstructural methods available for flood control, such as:

- Floodproofing;
- Flood warning mechanisms;
- Land-use controls such as zoning and development ordinances;
- Flood insurance programs;
- Flood preparedness activities;
- Public awareness and education programs;
- Existing primary and secondary drainage system maintenance;
- Acquisition of floodplain land or voluntary flood-prone property buyouts.

Nonstructural methods have gained attention in recent years as a viable approach to mitigating flood damage, despite being studied for over five decades. White initially proposed alternative approaches to structural methods in 1945. The Tennessee Valley Authority in the year 1950, successfully implemented a pilot program that focused on water management experiments. This program demonstrated the effectiveness of various nonstructural measures, yielding promising results. This led to increased interest from professionals and the American Institute of Civil Engineers, which played a role in creating a Manual for the Development of Flood Plain Rules and Regulations. In 1966, there was a significant push to incorporate nonstructural measures towards federal flood control policy (Yildirim et al., 2023).

In 1966, a Task Force was formed to evaluate the current federal flood prevention policy and propose improvements. Their focus was to establish a unified national plan to effectively manage and minimize flood losses. They identified five key objectives to guide this program. These were the measures and strategies implemented to manage and control floods in an academic context:

- Enhance the understanding of flood hazards at the local and regional levels;
- Enhance the planning and coordination of all advancements in the floodplain;
- Enhance the technical services provided to floodplain managers;
- Establish a comprehensive nationwide program for insurance against floods;
- Provide flexibility for future adjustments to federal flood prevention policy in response to evolving needs (Jamrussri & Toda, 2017).

5.2.2.1. Voluntary Buyouts of Flood-Prone Property

Acquiring high-risk floodplain lands from trustworthy sellers and moving vulnerable buildings and structures away from the floodplain are non-structural options that have gained more focus in recent times.

The devastating flood that occurred in the Midwest in 1993, affecting the Mississippi and Missouri rivers, and the subsequent studies and suggestions, generated widespread national interest in this specific solution. The key components of the optional buyout scheme are explicitly outlined in Higher Ground (Kraan et al., 2021).

The overarching objective of this option is theoretically straightforward but exceedingly challenging to implement. In theory, allocating public monies for a single outlay would lead to the relocation of a designated floodplain area. Despite its substantial expense, this investment would lead to the total eradication of flood hazards in certain regions. Ideally, the initial expense would be outweighed by the long-term financial gains, which would include lower flood losses and increased environmental benefits (De Vries & Fraser, 2012).

Higher ground outlines the optimal objectives of a carefully planned voluntary buyout initiative. These actions involve:

- Acquire a mixture of federal, state, and municipal monies to purchase high-risk assets from willing sellers as a one-time measure;
- Ensure the restoration of the purchased land to its original floodplain condition;
- Enact legislation to prevent the allocation of future disaster aid to that area;
- Provide support to previous property tenants and owners to relocate to safer areas and away from potential danger. Additionally, assists in the relocation of homes and businesses to areas outside the floodplain (Zavar et al., 2023).

This program offers distinct benefits compared to structural flood mitigation options and various nonstructural methods. One of the major advantages is the long-term assistance it offers to vulnerable individuals and assets in times of calamity. Another advantage is that the successful execution of this strategy would remove the need for continuous reliance on public funds for disaster relief, ensuring that tax dollars are utilized most efficiently. Finally, the program would fully achieve the environmental advantages of the floodplains by reestablishing them back to their genuine ecological purposes and thus offering many of the previously mentioned benefits (Brandon, 2019).

Did You Know?

Alternative approaches to flood control, such as voluntary property buyouts in flood-prone areas, prioritize adapting land use and construction methods to mitigate flood damage, rather than modifying the natural flow of water.

Purchasing properties as a means of mitigating flood risks has its fair share of limitations and drawbacks. Buyout programs may have an impact on the local tax base as significant businesses and investors may be compelled to explore opportunities in other communities. Insufficient enforcement of regulations or the absence of a well-planned land management strategy following the buyouts could lead to renewed development pressure. This may necessitate a revision of local regulatory laws to once again permit development in floodplains (Mukherji et al., 2024). Several buy-out plans have faced criticism for potentially displacing low-income residents from a central area of a community, which could lead to the development of more upscale projects in the future. Finally, the successful implementation of a buyout plan, no matter its scope, necessitates significant expenses, time, and meticulous planning. Considering the substantial alterations in land-use patterns that buyouts entail, particularly in established community areas, legal, social, and economic challenges can arise. These challenges may steer the decision-making process towards a more traditional structural approach (Ruppert et al., 2018).

5.3. FLOOD DAMAGE AND NET BENEFIT ESTIMATION

Learning Objectives:

- Understand the different categories of flood damages and their estimation techniques.
- Understand the methods for calculating expected damages and integrating risk analysis.
- Know the risk-based design procedures for hydraulic structures in flood management.

5.3.1. Damage Relationships

Flood damage is commonly reported as direct damage to property, but it's important to note that this is just one of the five empirical categories of damage. These categories include intangible damages, secondary damages, direct damages, indirect damages, and uncertainty damages. Indirect damage can occur due to the disruption of business operations, the need to address hardships, redirect traffic, and other associated consequences. Secondary damages occur due to the negative impact on individuals or services that rely on the affected property or disrupted services. Non-monetary losses encompass factors such as the state of the environment, the overall welfare of society, and the appreciation of aesthetic aspects. Damage caused by flooding is often a result of the constant uncertainty surrounding this natural disaster (Dedryver et al., 2010).

Several methods have been employed to assess the extent of direct harm. Grigg utilized three different categories of techniques, including empirical depth-damage curves, aggregate formulas, and historical damage curves. One of the well-known aggregate formulas is proposed by James (1972):

$$C_D = K_D U M_S h A \quad (1)$$

In the equation, C_D represents the cost of flood damage for a specific flood event; K_D is the rate of flood damage per foot of flood depth per dollar for the market value of the structure; U represents the proportion of floodplain that is developed for urban use; h represents the average flood depth over the area that is inundated, measured in feet; M_S is the market value for the structure that is affected by flooding, measured in dollars per established acre; lastly, A represents the total area that is flooded, measured in acres.

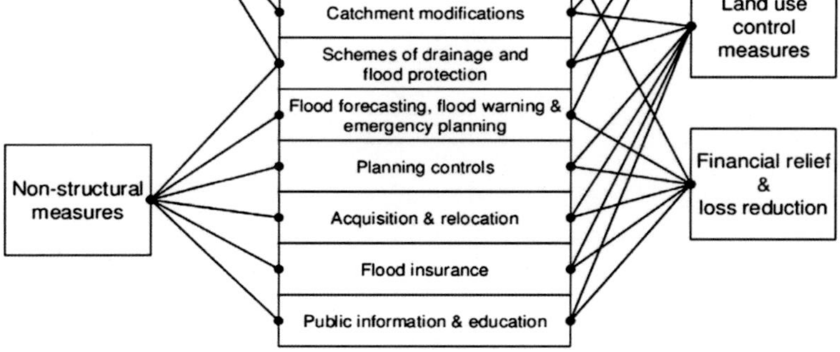

Figure 5.7. *Illustration of structural and non-structural measures.*

Source: Dagmar Haase; Creative Commons License.

Eckstein introduces a method called the historical damage curve, which involves plotting the damage caused by floods toward the flood stage (Jones et al., 2011). The utilization of empirical depth-damage curves (Figure 5.7) necessitates conducting a property survey for the floodplain and obtaining either an individual or combined estimation of depth (stage) instead of damaged curves for various elements such as crops, structures, utilities, roads, and more., that are located within the floodplain. The damage caused during this stage is then linked to the relationship between the stage discharges, resulting in the derivation of the damage-discharge relationship. This relationship is then combined with the discharge-frequency relationship to form the damage-frequency curve, as shown in Figure 5.8 (Irie et al., 2001).

5.3.2. Expected Damages

The annual predicted damage cost E (D) is calculated by determining the area within the damage-frequency curve, as illustrated in Figure 5.9(e), and can be mathematically expressed as:

$$E(D) = \int_{q_c}^{\infty} D(q_d) f(q_d) dq_d = \int_{q_c}^{\infty} D(q_d) dF(q_d) \tag{2}$$

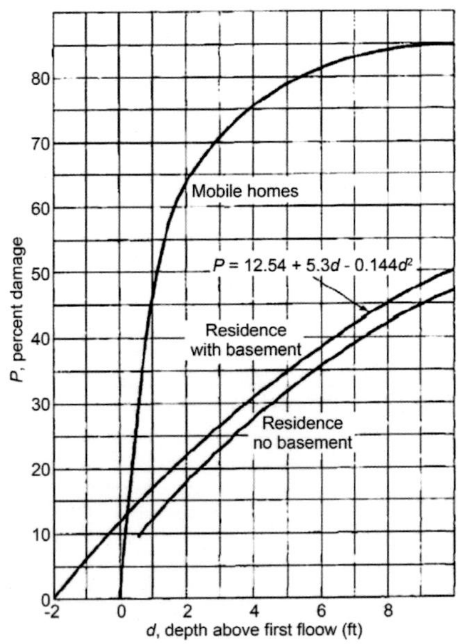

Figure 5.8. *Illustration of percent damage, mixed residences.*

Source: Corry; Creative Commons License.

When the threshold discharge q_c is exceeded, it can lead to potential damage. The flood damage D (q_d) varies for different discharges q_d, also known as the damage-discharge relationship, is characterized by the probability density function f (q_d) and F (q_d) along with the cumulative distribution function of discharge q_d. In real-world scenarios, the assessment of E (D) through Eqn. (2) is performed using numerical integration due to the intricate nature of damage functions and probability distribution functions (Eisenber & McDonnell, 2002).

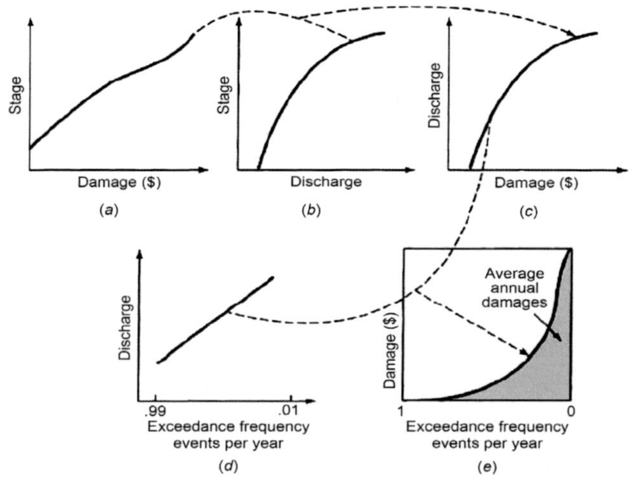

Figure 5.9. *Description of computation of average annual damages. (a) Stage-damage relation; (b) stage discharge; (c) discharge-damage; (d) discharge-frequency relation; and (e) damage-frequency relation.*

Source: Eleco Wan; Creative Commons License.

Thus, the shaded region in Figure 5.9(e) may be estimated using the trapezoidal rule as an illustration:

$$E(D) = \sum_{j=1}^{n} \frac{\left[D(q_j) + D(q_{j+1})\right]}{2} \left[F(q_{j+1}) - F(q_j)\right], \text{ for } q_c = q_1 \leq q_2 \leq \ldots \leq q_n < \infty \quad (3)$$

In which is the discretized discharge in the interval (q_c, ∞).

5.3.3. Risk-Based Analysis

Conventional design methods for hydraulic structures typically focus solely on the natural variability of hydrologic conditions. Typically, the likelihood of the hydraulic structure failing is assessed through probability analysis, as mentioned previously. Additional elements of hydrologic unpredictability are often overlooked. Techniques have been created to incorporate different elements of hydrologic unpredictability based on the risk design of hydraulic structures (Naidu et al., 2016).

Design approaches that consider risk integrate uncertainty assessment and reliability evaluation procedures. These approaches consider the financial trade-offs between the costs of the project and the anticipated costs of damage based on risk relationships. An optimization structure can incorporate a risk-based design process to determine the most optimal risk-based design. Thus, in an ideal risk-based design, the objective function considers expected annual damage along with the installation cost. The task at hand involves finding the most efficient structural sizes/capacities that result in the lowest total expected annual cost (TEAC) (Lucchi, 2016). From a mathematical perspective, the most efficient approach to design can be represented as:

$$\underset{\mathbf{x}}{\mathrm{Min}}\, TEAC = FC(\mathbf{x}) \cdot CRF + E(D|\mathbf{x}) \tag{4}$$

The total cost of installation (first cost) for the structure is denoted as FC. The capital recovery variable for converting the cost to an annual is represented by CRF. The vector for deciding variables related to structural sizes or capacities is denoted as 'x.' Lastly, the expected annual cost of damage caused by structural failure is represented as $E(D|x)$. Figure 5.10 displays a diagram that presents the cost computations over optimal risk-based design. Typically, calculating the initial cost is a simple process (Amland, 2000). The focus of risk-based design involves evaluating the cost of expected damage on an annual basis. The evaluation of $E(D|x)$varies based on the kinds of uncertainty that must be considered. Traditional approaches in water-resources engineering design focus solely on the natural variability of the hydrologic process, without considering other factors that may contribute to risk. Eqn. (2) can be used to calculate the mathematical illustration of the annual predicted damage under this condition (Sayers et al., 2002).

The method for calculations of expected yearly damages depicted in Figure 5.10 does not consider the uncertainty in the different relationships. Historically, uncertainties have been addressed by incorporating factors of safety, such as considering broad return periods or providing extra freeboard for levees. Recent advancements in risk analysis have enabled a more precise consideration of uncertainty (Sinha & Steel, 2015).

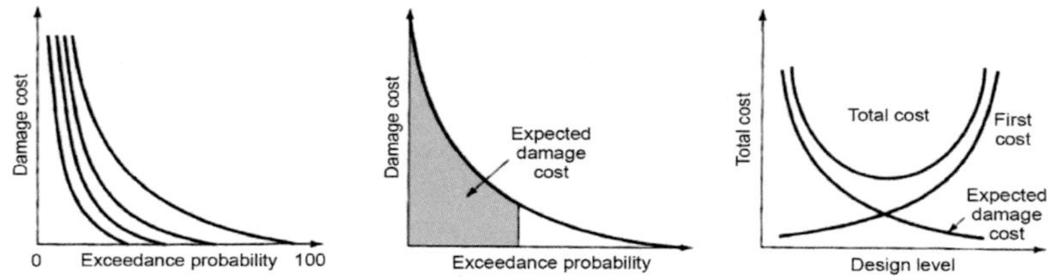

Figure 5.10. *Illustration of expected damage cost and total cost for risk-based design.*

Source: Corry; Creative Commons License.

Practice Problem

In the context of flood control measures, a municipality is evaluating two potential projects to mitigate the impact of recurrent floods in the region. Project A involves constructing levees along the riverbanks, while Project B proposes the restoration of natural wetlands to absorb excess floodwaters. To inform their decision, the municipality seeks to conduct a risk-based analysis for each project. For Project A, the cost of construction is estimated at $10 million, with an expected damage reduction of $15 million. Conversely, Project B entails an $8 million investment for wetland restoration, with an expected damage reduction of $12 million.

Solutions to Practice Problem

Conducting a risk-based analysis involves calculating the Benefit-Cost Ratio (BCR) for each project. Project A yields a BCR of 1.5, obtained by dividing the expected damage reduction by the cost of construction. Similarly, Project B's BCR is also 1.5, derived from the ratio of expected damage reduction to the cost of restoration. Subsequently, the net benefit of each project is determined by subtracting the total cost from the expected benefits. Project A yields a net benefit of $5 million, while Project B results in a net benefit of $4 million. Consequently, Project A emerges as the more cost-effective option, offering a higher net benefit for the investment made. This analysis displays how risk-based assessments and net benefit estimation aid in decision-making for flood control projects, enabling authorities to maximize effectiveness while considering financial constraints.

SUMMARY

- The Chapter, focus is on comprehensive strategies to mitigate the risks associated with flooding. It begins with a thorough exploration of floodplain management, starting with the definition and delineation of floodplains.
- Through hydrologic and hydraulic analyses, experts evaluate flood risk factors such as precipitation patterns and river discharge rates. This analysis informs the identification of floodways and their fringes, crucial for guiding development and preserving natural flood conveyance areas.
- Floodplain management strategies, including regulations and zoning policies, are then examined to minimize flood risk while promoting sustainable development. Additionally, the integration of stormwater management practices enhances resilience against flooding, particularly in urban areas where impervious surfaces exacerbate runoff.
- The chapter further delves into structural and nonstructural methods of flood control, highlighting approaches such as flood control reservoirs, channel modifications, and voluntary buyouts of flood-prone properties.
- Finally, it explores flood damage estimation, covering damage relationships, expected damages, and risk-based analysis to optimize flood control measures and minimize the economic impacts of flooding.

REVIEW QUESTIONS

1. What is the significance of floodplain management in mitigating flood risks?
2. Define floodplain and explain its importance in flood control measures.
3. How do hydrologic and hydraulic analyses aid in understanding and predicting floods?
4. Discuss the concept of floodways and floodway fringes and their role in floodplain management.
5. What are the key regulations governing floodplain management and how do they impact communities?
6. Differentiate between structural and nonstructural methods of flood control. Provide examples of each.

7. How do diversion channels, levees, and bypass channels contribute to flood damage reduction?
8. How are flood damages estimated, and what factors influence the estimation process?
9. Discuss the importance of risk-based analysis in evaluating flood damage and the benefits of control measures.

MULTIPLE CHOICE QUESTIONS

1. What does floodplain management primarily aim to achieve?

a. Increasing flood occurrences
b. Reducing flood risks and damages
c. Enhancing floodwater storage capacity
d. Promoting urbanization in flood-prone areas

2. What is the main purpose of conducting hydrologic and hydraulic analysis of floods?

a. To increase water flow during floods
b. To assess the impact of floods on ecosystems
c. To predict flood inundation areas and depths
d. To regulate floodway fringes

3. Which structural method is commonly used for flood damage reduction by storing excess water?

a. Flood control reservoirs
b. Diversion channels
c. Levees
d. By-pass channels

4. What is the primary purpose of diversion channels, levees, and bypass channels?

a. To increase floodwater storage capacity
b. To mitigate environmental impacts of floods
c. To redirect floodwaters away from populated areas
d. To enhance stormwater management

5. Which nonstructural method involves the voluntary purchase of flood-prone properties?

a. Structural reinforcements
b. Flood control reservoirs

c. Diversion channels

d. Voluntary buyouts

6. What is the primary focus of flood damage estimation?

a. Estimating financial losses only

b. Assessing environmental impacts of floods

c. Predicting flood occurrence probabilities

d. Evaluating the extent of property damage and loss of life

Answers to Multiple Questions

1. (b); 2. (c); 3. (a); 4. (c); 5. (d); 6. (d)

REFERENCES

1. Aksoy, H., Kirca, V. S. O., Burgan, H. I., & Kellecioglu, D. (2016). Hydrological and hydraulic models for determination of flood-prone and flood inundation areas. *Proceedings of the International Association of Hydrological Sciences, 373*, 137–141. https://doi.org/10.5194/piahs-373-137-2016.
2. Amland, S. (2000). Risk-based testing: Risk analysis fundamentals and metrics for software testing including a financial application case study. *Journal of Systems and Software, 53*(3), 287–295. https://doi.org/10.1016/S0164-1212(00)00016-9.
3. Anselmo, V., Galeati, G., Palmieri, S., Rossi, U., & Todini, E. (1996). Flood risk assessment using an integrated hydrological and hydraulic modeling approach: A case study. *Journal of Hydrology, 175*(1–4), 533–554. https://doi.org/10.1016/S0022-1694(96)80023-0.
4. Aureli, F., Mignosa, P., Prost, F., & Dazzi, S. (2021). Hydrological and hydraulic flood hazard modeling in poorly gauged catchments: An analysis in Northern Italy. *Hydrology, 8*(4), 149. https://doi.org/10.3390/hydrology8040149.
5. Bouchakour, M., Zhao, X., Ge, J., Miclău ,C., & Yang, B. (2022). Evolution of submarine channel morphology in intra-slope mini-basins: 3D-seismic interpretation from offshore Niger delta. *Marine and Petroleum Geology, 146*, 105912. https://doi.org/10.1016/j.marpetgeo.2022.105912.
6. Braden, J. B., & Johnston, D. M. (2004). Downstream economic benefits from storm-water management. *Journal of Water Resources Planning and Management, 130*(6), 498–505. https://doi.org/10.1061/(ASCE)0733-9496(2004)130:6(498).

7. Brandon, T. (2019). Sea level rise planning for socially vulnerable communities: A more equitable approach to federal buyout programs. *U. Det. Mercy L. Rev., 97*, 435.

8. Burby, R. J. (2001). Flood insurance and floodplain management: The US experience. *Global Environmental Change Part B: Environmental Hazards, 3*(3), 111–122. https://doi.org/10.1016/S1464-2867(02)00003-7.

9. Carver, M. B. (1969, September). Effect of by-pass characteristics on parallel-channel flow instabilities. In *Proceedings of the Institution of Mechanical Engineers, Conference Proceedings* (Vol. 184, No. 3, pp. 84–92). https://doi.org/10.1243/PIME_CONF_1969_184_104_02.

10. Cheng, C. T., & Chau, K. W. (2004). Flood control management system for reservoirs. *Environmental Modelling & Software, 19*(12), 1141–1150. https://doi.org/10.1016/j.envsoft.2004.05.001.

11. Croke, J., Fryirs, K., & Thompson, C. (2016). Defining the floodplain in hydrologically-variable settings: Implications for flood risk management. *Earth Surface Processes and Landforms, 41*(14), 2153–2164. https://doi.org/10.1002/esp.3962.

12. Davar, K. S., Henderson, J. M., & Burrell, B. C. (2001). Flood damage reduction. *Water International, 26*(2), 162–176. https://doi.org/10.1080/02508060108686908.

13. De Vries, D. H., & Fraser, J. C. (2012). Citizenship rights and voluntary decision making in post-disaster US floodplain buyout mitigation programs. *International Journal of Mass Emergencies & Disasters, 30*(1), 1–33.

14. Dedryver, C. A., Le Ralec, A., & Fabre, F. (2010). The conflicting relationships between aphids and men: A review of aphid damage and control strategies. *Comptes Rendus Biologies, 333*(6–7), 539–553. https://doi.org/10.1016/j.crvi.2010.03.009.

15. Dogruoz, M. B., Urdaneta, M., & Ortega, A. (2005). Experiments and modeling of the hydraulic resistance and heat transfer of in-line square pin fin heat sink with top by-pass flow. *International Journal of Heat and Mass Transfer, 48*(23–24), 5058–5071. https://doi.org/10.1016/j.ijheatmasstransfer.2005.07.001.

16. Droz, L., Jégou, I., Gillet, H., Dennielou, B., Bez, M., Canals, M., & Rabineau, M. (2020). On the termination of deep-sea fan channels: Examples from the Rhône Fan (Gulf of Lion, Western Mediterranean Sea). *Geomorphology, 369*, 107368. https://doi.org/10.1016/j.geomorph.2020.107368.

17. Eisenberg, M. A., & McDonnell, B. H. (2002). Expectation damages and the theory of overreliance. *Hastings LJ, 54*, 1335.
18. Elosegi, A., & Sabater, S. (2013). Effects of hydro morphological impacts on river ecosystem functioning: A review and suggestions for assessing ecological impacts. *Hydrobiologia, 712*, 129–143. https://doi.org/10.1007/s10750-012-1226-6.
19. Faisal, I. M., Kabir, M. R., & Nishat, A. (1999). Non-structural flood mitigation measures for Dhaka City. *Urban Water, 1*(2), 145–153. https://doi.org/10.1016/S1462-0758(00)00014-0.
20. Gotoh, H., Maeno, Y., Takezawa, M., & Ohnishi, M. (2011). Flood control and small-scale reservoirs. *River Basin Management VI, 146*, 51–60.
21. Gül, G. O., Harmancıoğlu, N., & Gül, A. (2010). A combined hydrologic and hydraulic modeling approach for testing efficiency of structural flood control measures. *Natural Hazards, 54*, 245–260. https://doi.org/10.1007/s11069-009-9465-0.
22. Haris, H., Chow, M. F., Usman, F., Sidek, L. M., Roseli, Z. A., & Norlida, M. D. (2016). Urban stormwater management model and tools for designing stormwater management of green infrastructure practices. In *IOP Conference Series: Earth and Environmental Science* (Vol. 32, No. 1, p. 012022). https://doi.org/10.1088/1755-1315/32/1/012022.
23. Hayes, B. D. (2004). Interdisciplinary planning of nonstructural flood hazard mitigation. *Journal of Water Resources Planning and Management, 130*(1), 15–25. https://doi.org/10.1061/(ASCE)0733-9496(2004)130:1(15).
24. Hornung, L. K., Podschun, S. A., & Pusch, M. (2019). Linking ecosystem services and measures in river and floodplain management. *Ecosystems and People, 15*(1), 214–231. https://doi.org/10.1080/26395916.2019.1667876.
25. Hosseinzadeh, M. M., Khaleghi, S., & Panahi, R. (2022). Determining the floodway and flood fringe of the Gamasiab River, Kermanshah, Iran. *International Journal of Water, 15*(2), 117–130. https://doi.org/10.1504/IJW.2022.10048833.
26. Hupp, C. R., Pierce, A. R., & Noe, G. B. (2009). Floodplain geomorphic processes and environmental impacts of human alteration along coastal plain rivers, USA. *Wetlands, 29*, 413–429. https://doi.org/10.1672/08-169.1.
27. Irie, M., Asami, S., Nagata, S., Miyata, M., & Kasai, H. (2001). Relationships between perceived workload, stress, and oxidative DNA damage. *International Archives of Occupational and Environmental Health, 74*, 153–157. https://doi.org/10.1007/s004200000198.

28. James, L. D. (1965). Nonstructural measures for flood control. *Water Resources Research, 1*(1), 9–24. https://doi.org/10.1029/WR001i001p00009.
29. Jamrussri, S., & Toda, Y. (2017). Simulating past severe flood events to evaluate the effectiveness of nonstructural flood countermeasures in the upper Chao Phraya River Basin, Thailand. *Journal of Hydrology: Regional Studies, 10*, 82–94. https://doi.org/10.1016/j.ejrh.2016.12.084.
30. Jolly, B. A., Anyiam, O. A., & Omeru, T. (2017). Structural controls on channel-related seismic facies distribution in the toe-thrust of deepwater Niger Delta. *Journal of African Earth Sciences, 125*, 151–165. https://doi.org/10.1016/j.jafrearsci.2016.11.022.
31. Jones, T., Dacin, P. A., & Taylor, S. F. (2011). Relational damage and relationship repair: A new look at transgressions in service relationships. *Journal of Service Research, 14*(3), 318–339. https://doi.org/10.1177/1094670511404396.
32. Keller, E. A. (2020). Channelization: environmental, geomorphic, and engineering aspects. In *Geomorphology and Engineering* (pp. 115–140). https://doi.org/10.1201/9781003078075-6.
33. Kiedrzyńska, E., Kiedrzyński, M., & Zalewski, M. (2015). Sustainable floodplain management for flood prevention and water quality improvement. *Natural Hazards, 76*, 955–977. https://doi.org/10.1007/s11069-014-1535-3.
34. Kraan, C. M., Hino, M., Niemann, J., Siders, A. R., & Mach, K. J. (2021). Promoting equity in retreat through voluntary property buyout programs. *Journal of Environmental Studies and Sciences, 11*(3), 481–492. https://doi.org/10.1007/s13412-021-00687-8.
35. Kreibich, H., Bubeck, P., Van Vliet, M., & De Moel, H. (2015). A review of damage-reducing measures to manage fluvial flood risks in a changing climate. *Mitigation and Adaptation Strategies for Global Change, 20*, 967–989. https://doi.org/10.1007/s11027-014-9629-5.
36. Krizek, M., & Sobr, J. W. (2006). Floodplain and its delimitation. *Geografie--Sbornik ceske geograficke spolecnosti, 111*(3), 260–273.
37. Li, C., Cheng, X., Li, N., Du, X., Yu, Q., & Kan, G. (2016). A framework for flood risk analysis and benefit assessment of flood control measures in urban areas. *International Journal of Environmental Research and Public Health, 13*(8), 787. https://doi.org/10.3390/ijerph13080787.
38. Liebman, E. (1973). Legal problems in regulating flood hazard areas. *Journal of the Hydraulics Division, 99*(11), 2113–2123.

39. Lucchi, E. (2016). Multidisciplinary risk-based analysis for supporting the decision-making process on conservation, energy efficiency, and human comfort in museum buildings. *Journal of Cultural Heritage, 22*, 1079–1089. https://doi.org/10.1016/j.culher.2016.06.006.

40. Mansour, M. F., Morgenstern, N. R., & Martin, C. D. (2011). Expected damage from displacement of slow-moving slides. *Landslides, 8*, 117–131. https://doi.org/10.1007/s10346-010-0220-0.

41. Meyer, V., Priest, S., & Kuhlicke, C. (2012). Economic evaluation of structural and non-structural flood risk management measures: Examples from the Mulde River. *Natural Hazards, 62*, 301–324. https://doi.org/10.1007/s11069-011-9994-1.

42. Miguez, M. G., & de Magalhães, L. P. C. (2010). Urban flood control, simulation, and management–An integrated approach. In A. C. de Pina Filho & A. C. de Pina (Eds.), *Methods and Techniques in Urban Engineering* (pp. 131–160). InTech. https://doi.org/10.5772/13313.

43. Mukherji, A., Adeniji, K. N., Curtis, S., Helgeson, J., Kruse, J., & Millea, M. (2024). Buyout program experiences and perspectives of local public officials in eastern North Carolina. *Environmental Hazards, 2*(1), 1–19. https://doi.org/10.1080/02723638.2023.1875187.

44. Naidu, R., Espana, V. A. A., Liu, Y., & Jit, J. (2016). Emerging contaminants in the environment: Risk-based analysis for better management. *Chemosphere, 154*, 350–357. https://doi.org/10.1016/j.chemosphere.2016.03.068.

45. Newcomer Johnson, T. A., Kaushal, S. S., Mayer, P. M., & Grese, M. M. (2014). Effects of stormwater management and stream restoration on watershed nitrogen retention. *Biogeochemistry, 121*, 81–106. https://doi.org/10.1007/s10533-014-9969-0.

46. Olsen, J. R. (2006). Climate change and floodplain management in the United States. *Climatic Change, 76*(3), 407–426. https://doi.org/10.1007/s10584-005-9020-3.

47. Prékopa, A., & Szántai, T. (1978). Flood control reservoir system design using stochastic programming. *Mathematical Programming in Use, 4*(2), 138–151. https://doi.org/10.1007/BFb0007843.

48. Rahmani-Rezaeieh, A., Mohammadi, M., & Danandeh Mehr, A. (2020). Climate change impacts on floodway and floodway fringe: A case study in Shahrchay River Basin, Iran. *Arabian Journal of Geosciences, 13*(12), 494. https://doi.org/10.1007/s12517-020-05478-7.

49. Ren, M., He, X., Kan, G., Wang, F., Zhang, H., Li, H., & Zhang, Z. (2017). A comparison of flood control standards for reservoir engineering for

different countries. *Water, 9*(3), 152. https://doi.org/10.3390/w9030152.

50. Ruppert, T., Fergus, J., & Russe-Garcia, E. (2018). Managing property buyouts at the local level: Seeking benefits and limiting harms. *Environmental Law Reporter News & Analysis, 48*, 10520.
51. Sayers, P. B., Hall, J. W., & Meadowcroft, I. C. (2002, May). Towards risk-based flood hazard management in the UK. In *Proceedings of the Institution of Civil Engineers-Civil Engineering* (Vol. 150, No. 5, pp. 36–42). https://doi.org/10.1680/cien.2002.150.5.36.
52. Schoof, R. (1980). Environmental impact of channel modification. *JAWRA Journal of the American Water Resources Association, 16*(4), 697–701. https://doi.org/10.1111/j.1752-1688.1980.tb02504.x.
53. Sinha, Y., & Steel, J. A. (2015). A progressive study into offshore wind farm maintenance optimization using risk-based failure analysis. *Renewable and Sustainable Energy Reviews, 42*, 735–742. https://doi.org/10.1016/j.rser.2014.10.090.
54. Smemoe, C. M., Nelson, E. J., Zundel, A. K., & Miller, A. W. (2007). Demonstrating floodplain uncertainty using flood probability maps. *JAWRA Journal of the American Water Resources Association, 43*(2), 359–371. https://doi.org/10.1111/j.1752-1688.2007.00028.x.
55. Surian, N., Ziliani, L., Comiti, F., & Lenzi, M. A. (2009). Channel adjustments and alteration of sediment fluxes in gravel-bed rivers of North-Eastern Italy: Potentials and limitations for channel recovery. *River Research and Applications, 25*(5), 551–567. https://doi.org/10.1002/rra.1208.
56. Thampapillai, D. J., & Musgrave, W. F. (1985). Flood damage mitigation: A review of structural and nonstructural measures and alternative decision frameworks. *Water Resources Research, 21*(4), 411–424. https://doi.org/10.1029/WR021i004p00411.
57. Vangelis, H., Zotou, I., Kourtis, I. M., Bellos, V., & Tsihrintzis, V. A. (2022). Relationship of rainfall and flood return periods through hydrologic and hydraulic modeling. *Water, 14*(22), 3618. https://doi.org/10.3390/w14223618.
58. Walesh, S. G. (1979). Flood insurance program: Regional agency's perspective. *Journal of the Water Resources Planning and Management Division, 105*(2), 243–257.
59. Yildirim, E., Alabbad, Y., & Demir, I. (2023). Non-structural flood mitigation optimization at community scale: Middle Cedar Case Study. *Journal of Environmental Management, 346*, 119025. https://doi.org/10.1016/j.jenvman.2020.119025.

60. Zavar, E., Binder, S. B., Greer, A., & Breaux, A. (2023). Using the past to understand future property acquisitions: An examination of historic voluntary and mandatory household relocations. *Natural Hazards, 116*(2), 1973–1993. https://doi.org/10.1007/s11069-021-04841-8.

CHAPTER 6

FLOOD FORECASTING, WARNING, AND FLOOD FIGHTING

LEARNING OBJECTIVES

At the end of this chapter, readers will be able to:

1. Understand the significance of flood forecasting and warning for economic advancement.
2. Explain the importance of historical hydrological data and current information in developing flood-forecasting procedures.
3. Know the key components and techniques involved in flood forecasting, including rainfall-runoff relationships and unit hydrographs.
4. Describe the methods used for predicting stages in large streams and the application of the Peak Travel Time Method.
5. Recognize the role of flood warning signals and emergency engineering measures in flood control and management.
6. Understand the engineering measures for protection against sliding, overtopping, scouring, and closure of breaches in embankments during floods.
7. Analyze the process of long-term forecasting limitations and its significance in water resources management.

INTRODUCTORY EXAMPLE

In the wake of increasing climate variability, RiverTech, a prominent environmental consultancy firm, partners with the Greenwood County Government to tackle the recurrent threat of flooding in the region. With its expertise in hydrology and engineering, RiverTech aids Greenwood County in implementing effective flood forecasting, warning, and mitigation measures.

Following heavy rainfall forecasts, RiverTech utilizes sophisticated forecasting techniques and historical data analysis to predict potential flood occurrences along the Greenwood River. By understanding the hydrological dynamics of the watershed, they can estimate the timing, magnitude, and duration of flood events with greater accuracy.

Armed with this information, Greenwood County activates its flood warning system, promptly alerting residents and emergency responders to impending flood risks. Meanwhile, RiverTech collaborates with the county's engineering department to implement practical flood fighting measures, such as reinforcing levees, installing flood gates, and establishing emergency evacuation routes.

As floodwaters rise, the collaborative efforts of RiverTech and Greenwood County prove instrumental in minimizing the impact on communities and infrastructure. Through proactive flood management strategies and timely interventions, they exemplify the importance of effective partnership and preparedness in mitigating the devastating effects of natural disasters like floods.

UNIT INTRODUCTION

Flood forecasting involves predicting the future of water levels, extents, and depths of flooding in rivers and floodplains. Flood warning involves the creation of forecasts presented in a comprehensible manner. Hydrological forecasts, together with meteorological forecasts, hold significant importance in nearly all countries worldwide, particularly for their economic progress. Forecasts are highly valuable in water resources engineering for optimizing the use of river energy, inland transportation, irrigation, water supply, and flood management (Parker & Fordham, 1996).

Accurate predictions of potentially catastrophic events like river floods are crucial for the well-being of a country's economy and the safety of its rural and urban communities residing in flood-prone areas. By providing timely warnings in advance, it becomes possible to minimize the extent of damage by implementing precautionary measures such as evacuating individuals and property to safer locations. Efficient operation of existing reservoirs greatly relies on accurate flood forecasting (Jain et al., 2018). When a flood is predicted, the reservoir can be appropriately managed to mitigate its effects and maintain sufficient storage for emergency flood relief. The first high reservoir level can be drained to the lowest level before the flood wave arrives if a flood forecasting service is available. This will provide enough storage for the flood's moderation and any necessary adjustments to the reservoir's operation to pass the flood while also having the necessary storage at the end (Brilly & Polic, 2005).

Sufficient data is required for the development of the forecasting service. To develop the river forecasting technique, historical hydrological data is necessary. Additionally, adequate current information is required for the generation of operational forecasts. Typically, it is essential to possess at least 10 years of fundamental hydraulic data to establish effective river forecasting methods. The key criterion is that the data should cover a realistic range of peak flows. Sufficient rainfall data should be available to accurately estimate the amount of precipitation in the area being analyzed (Emerton et al., 2016).

The amount of rainfall data needed can vary depending on the shape of the basin and the weather conditions. For example, regions with inconsistent rainfall patterns require more advanced techniques compared to areas with consistent rainfall. To develop an effective forecast system for stream flow, it is crucial for the forecaster to have accurate and dependable information about the hydrological conditions in the basin. This includes details about the amount and distribution of rainfall, as well as the water equivalent program of snow if it is present (Borga et al., 2011). More details regarding the snow conditions and duration of showers are required. Additionally, it is necessary to have reports on the river stage at crucial locations, since they are essential for both predicting future conditions and confirming the accuracy of the data utilized in these predictions (Chau et al., 2005). Data must flow quickly to prevent warning time loss. It has been discovered that telephonic and telegraphic contact can occasionally break down during storms and is therefore not regarded as reliable. Installing a wireless network to link all the observation sites to the forecast control room is therefore recommended. Plain and hilly areas usually use very high-frequency links (82.9 megacycles) and high-frequency links (6.9545 megacycles during the day and 3.954 megacycles at night) (Chang et al., 2007).

The VHF sets are powered by 12 V batteries, while the HF sets are powered by 230 V AC. Therefore, it is essential to have main-operated battery chargers available at specific locations to recharge the batteries. Additionally, diesel engine generators are typically provided as a backup for 230 V AC sets. A complete communication network is designed to connect the central control room with the rain gauges and discharge stations of each major river system (Tariq et al., 2012).

6.1. FORECASTING PROCEDURES AND TECHNIQUES

Learning Objectives:

- Understand river forecasting techniques: unit hydrographs, rainfall-runoff, and stage-discharge relations.

This section covers the common and conventional river forecasting ideas that are encountered in daily practice. The rainfall-runoff connection, unit hydrographs, stage discharge relationships, synthetic unit hydrographs, flood routing techniques, etc., are some of the fundamental approaches utilized in river forecasting (Kunst & Neusser, 1986). Basically, forecasting consists of two primary parts:

- Forecasting the magnitude of runoff generated by storm rainfall; and
- Forecasting the flow of water over time at predetermined points along the river. During the practical procedure, it involves estimating:
  - The level of water in the stream;
  - The timing of the stage and the corresponding peak discharges;
  - The peak discharge magnitude;
  - The volume of the flood runoff.

Therefore, it will be seen that forecasting encompasses the following:

- Determining runoff from rainfall data;
- Predicting discharge stages based on the control of runoff flow and transportation in the river channel and computing the flood hydrograph using runoff volumes (Harvey, 1984).

Streamlining the forecasting organizations and procedures is essential since the forecast needs to be created as quickly as possible. Special forms must be used for this purpose to log incoming reports, enter them into rainfall plotting maps, and create continuous hydrograph charts. Furthermore, correlation charts that show the relationship between rainfall and runoff while accounting for antecedent precipitation index, season, and other variables should be created using historical data. Forecasting sheets need to be prepared ahead of time, dated and leveled (Bodily & Freeland, 1988).

When dealing with areas close to the beginning of the river, it is typically enough to predict the maximum height and timing of the flood. Considering the quick fluctuations in water levels, the limited time spent above the flood stage does not contribute much to accurately forecasting the entire hydrograph. When it comes to these situations, relying solely on empirical charts that show the relationship between flow at a specific location and stream flow events at an upstream station, as well as the connection between rainfalls and crest-stage, is sufficient for making accurate forecasts. When dealing with stations located in the lower sections of large rivers, where the rate of rise is gradual, it becomes crucial to predict the timing of reaching different critical heights during both the rising and falling stages (Chatfield, 1978).

In this case, basin runoff routing, unit hydrographs, and flood wave routing techniques will be utilized. In order to accurately forecast floods, it is necessary to determine the entire hydrograph at the upper stations for downstream forecasting. This is crucial for the effective operation of flood control reservoirs, as it allows for the prediction of maximum storage space to reduce flood intensity. Additionally, it ensures that the reservoirs are filled by the end of the flood period (Mentzer & Kahn, 1995).

If there is enough data available, the unit hydrograph method can be used for flood forecasting. The rainfall-runoff relationship can be used to estimate the quantity of water that is projected to flow in the stream. This is done by using the unit hydrograph and stream-flow routing algorithms, which help identify how the water is distributed over time at a specific forecast point. Stage discharge relationships are subsequently utilized to establish a connection between the flow and the stages (Milkovich et al., 1972).

6.2. RUNOFF FROM RAINFALL DATA DETERMINATION

Learning Objectives:

- Understand the application of rainfall-runoff correlation using graphical techniques.

If there is enough data available, the unit hydrograph method can be used for flood forecasting. The rainfall-runoff relationship can be used to estimate the quantity of water that is projected to flow in the stream. This is done by using the unit hydrograph and stream-flow routing algorithms, which help identify how the water is distributed over time at a specific forecast point. Stage discharge relationships are subsequently utilized to establish a connection between the flow and the stages (Horner & Jens, 1942).

When it comes to flood forecasting, it is important to consider various factors such as storm duration, antecedent basin conditions, storm rainfall, and the resulting runoff. Usually, these variables are measured and reported as an average depth over the basin in millimeters (Skaugen & Onof, 2014). One of the techniques commonly used is the coaxial graphical method, as shown in Figure 6.1. In this study, the antecedent basin conditions using two variables: the antecedent precipitation index (API) and the week of the year will be examined. The API, which serves as an indicator of moisture conditions within the basin, is essentially a calculation that takes into account the amount of precipitation that occurred before the storm, with different weights assigned based on when it occurred (Hawkins, 1993).

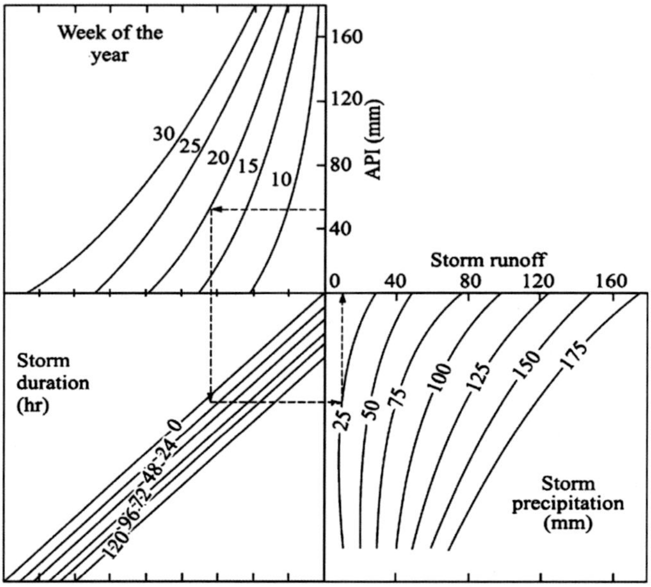

Figure 6.1. *Illustration of co-axial rainfall runoff correlation.*

Source: Gareth Pender; Creative Commons License.

An equation of this type is typically defined as:

$$P_a = k^1 p_1 + k^2 p_2 \ldots . + k^t p_t$$

where; P_a is the antecedent precipitation index; is the total quantity of precipitation that occurred 't' days before the storm in question; and is the recession factor (Dorum et al., 2010).

Therefore, the index for any given day is calculated by multiplying the index of the previous day by a constant factor, k. When rain falls, the index after the rainfall is determined by adding the amount of rainfall on that day to the index before the rain (Soulis & Valiantzas, 2013). The impact of the rainfall that occurs 20–60 days before the storm being studied is typically accounted for when calculating the API, based on the desired level of accuracy for the specific location. The range of values for k is between 0.85 and 0.93. During a storm, the characteristics of the basin's interception and evapotranspiration, along with the API, can provide valuable information about the antecedent soil conditions (Soulis & Valiantzas, 2012). Once the amount of runoff has been determined for a specific storm scenario, it is important to further calculate how this water is distributed over time at the forecast location. The unit hydrograph is a simple but highly dependable instrument in this task. Time unit hydrographs with durations of 6–12 hours are utilized to address the issue of unequal distribution of runoff in short periods (Blume, 2007). Estimations are made for the increase in runoff during each time period, taking into account the combined contribution from each interval on top of the previous ones. After determining the flood hydrograph at a particular place on a river, the subsequent task is to predict the flood hydrograph at points downstream in order to make predictions. Any of the routing strategies outlined above can be used to do this (Kisi et al., 2013).

6.3. FORECASTING STAGES METHODS

Learning Objectives:

- Understand peak travel time method for river stage prediction.

This section discusses methods for estimating the stage for big streams. Assuming there are no substantial tributaries or local inputs from tiny streams (Dazzi et al., 2021), the stage is known at the upstream station and must be computed or expected at the downstream station. The process of determining the stage at a downstream station with respect to the stage upstream is a basic kind of river forecasting (Hao & Tian, 2019).

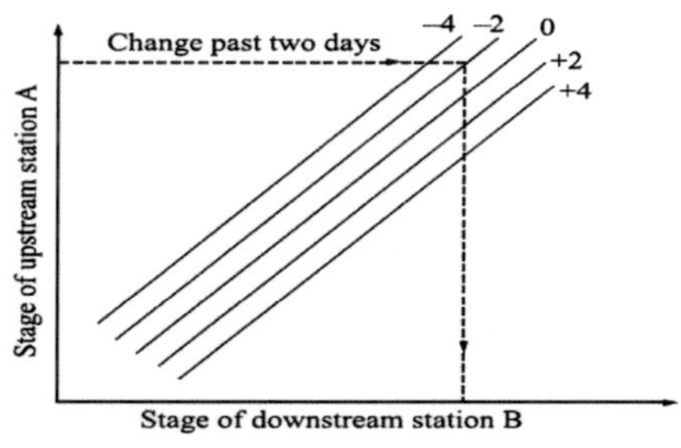

Figure 6.2. *Description of stage at upstream station vs. stage at downstream station.*

Source: Hazel Faulkner; Creative Commons License.

By analyzing historical stream flow records, it is possible to create a set of curves that illustrate the relationship between the stage of the downstream station and the stage of the upstream station (Han & Coulibaly, 2017). A third parameter, the number of days that have passed since the stage began, is introduced by these sets of curves (Figure 6.2). This parameter shows how long it takes the flood hydrograph to cross the river's reach (Wang et al., 2018).

6.3.1. Peak Travel Time Method

The duration for a peak to move between two stations is determined by the travel time of the river stretch. The journey time is crucial for predicting the stage. The fluctuation of the variable is influenced by the stages and channel conditions, rather than staying constant. As the river level nears the top of the banks, lower river stages tend to be briefer, resulting in shorter transit times. When the river reaches a high enough level to flood nearby areas and form overbank stages, the time it takes for the water to flow may start to climb again. This is because the overbank stages have rougher surfaces compared to the main river channel. Figure 6.3 shows the relationship between journey time and the stage in the river (Higatani et al., 2009).

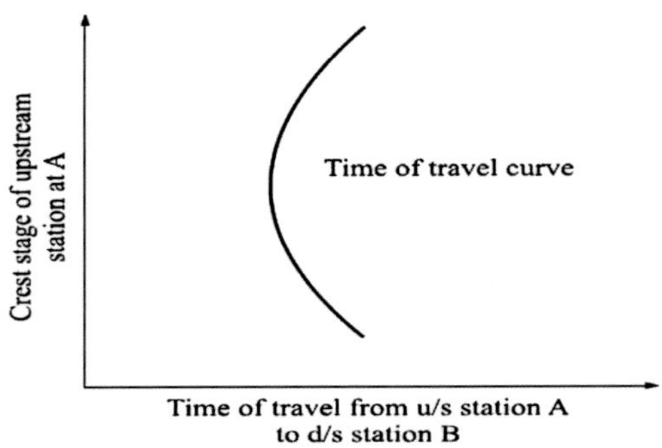

Figure 6.3. *Illustration of variation of time of travel with stage in the river.*

Source: Jochen Schanze; Creative Commons License.

Based on observations, the peak moves at a speed between 1.4 and 2.0 times the average water velocity. It is also feasible to establish a link based on the river's attributes. If the streams' hydraulic radii and roughness are comparable, the time of travel can be calculated as follows (Ban et al., 2010).

$$V_1 = C\sqrt{m_1 s_1}\, V_2 = C\sqrt{m_2 s_2}$$

where; the reaches' respective velocities are denoted by V_1 and V_2; the hydraulic mean radii are denoted bym$_1$ and m_2; the slopes of the reaches are denoted by s_1 and s_2 and *C* is Chezy's constant.

Consequently, the journey time:

$$t_1 = \frac{L_1}{L_2} \ and \ t_2 = \frac{L_2}{V_2}$$

$$t_2 = \frac{L_2}{V_2} t_1 \sqrt{{s_1}/{s_2}}$$

where; the first and second reaches' respective travel times are denoted by t_1 and t_2; their corresponding slopes by s_1 and s_2; and their corresponding lengths by L_1 and L_2. In addition to the detailed discussion on short-term forecasting, the issue of long-term forecasting holds immense importance in effectively managing water resources, like other natural resources. However, finding a solution relies heavily on addressing the challenge of accurately predicting long-term precipitation (Zou et al., 2014). Accurate rainfall prediction holds great significance in regions such as India, characterized by a tropical and monsoon climate, where rainfall is the primary contributor to runoff (Brauchler et al., 2003). At present, the understanding of meteorological science has not advanced enough to provide accurate long-term predictions of rainfall. The potential for long-term runoff forecasting success is limited (Alemazkoor et al., 2015).

6.3.2. Forecasts Reporting Example

In the following section, an easy process for preparing a flood forecast for a small drainage basin will be discussed. Figure 6.4 shows a hypothetical river basin where forecasting needs to be conducted at stations A and B. Unit hydrographs are often used to distribute runoff, while the Muskingum routing technique is utilized for flood routing from stations A and B (Cassar & Gibson, 2008).

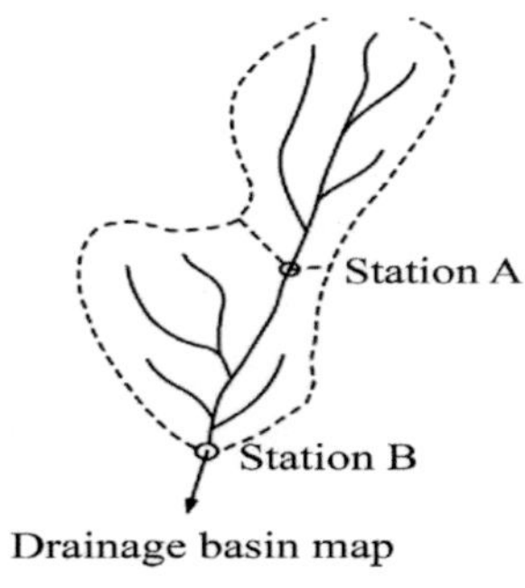

Figure 6.4. *Illustration of the drainage basin.*

Source: Even Zeman; Creative Commons License.

6.4. FLOOD WARNING

Learning Objectives:

- Differentiate between protected and unprotected floodplain areas for warning.
- Understand the signaling system for flood warnings: yellow, amber, red.

The flood warning authority, typically the engineer in charge, classifies the vulnerable areas into two main categories: unprotected and protected areas. Protected areas refer to the sections of flood plains that have been protected from flooding through the construction of embankments along riverbanks (Drobot & Parker, 2007). Other areas within the flood plain that are at risk of flooding are currently without any form of protection. Typically, the at-risk areas are organized around flood warning stations. Each station designates a specific zone where warnings are issued when the river hits the warning stage at that station (Kuller et al., 2021).

Typically, flood warnings are communicated using various types of signals:

- **Yellow Signal:** This provides early warning of an approaching flood in the specific area of the river.
- **Amber Signal:** This signal serves as an early warning for an approaching high flood, leading the departmental staff to heighten their vigilance and implement necessary precautions. Warning should be exercised in areas without proper protection (Thieken et al., 2023).
- **Red Signal:** This suggests that a significant flood is approaching, which has the potential to cause harm to the embankment in protected regions and result in flooding of at least 1.0 meters beyond the normal level of agricultural land in unprotected areas (Parker & Handmer, 1998).

The yellow warning is intended for internal use and should not be shared with the public. It is specifically meant for departmental authorities and staff responsible for vital installations, including District Magistrate, General Administration, Block Development Officers, Sub-divisional Officers, Health, Civil Defense, Police, Public Works Department, Electricity Boards, Food and Supply, Oil Storage, Air Force, Army, Railways, Telegraphs and Posts, Water Supply, and Installations (Carsell, 2004).

Practice Problem

As part of the emergency management team at ATLANTECH disaster association, you are responsible for monitoring flood risks along a river that borders the company's facilities. Today, the monitoring station reports a current river stage of 12.5 meters. Your team has established threshold levels for issuing flood warnings: a warning stage threshold at 15 meters, an amber signal at 17 meters, and a red signal at 19 meters. Given the current river stage, your first task is to assess if any warning signals should be issued to alert company personnel and neighboring communities. Next, you need to calculate the difference between the current river stage and the warning stage threshold to gauge the severity of the situation. Finally, considering the steady rise of 0.5 meters per hour, you must estimate the time until the red signal threshold is reached. By addressing these challenges, you will ensure the safety of employees, protect company assets, and maintain business continuity amidst potential flooding risks.

Solutions to Practice Problem

After evaluating the current river stage of 12.5 meters, it falls below the established warning stage threshold of 15 meters, indicating no immediate need for issuing warning signals. However, there remains a significant difference of 2.5 meters between the current river stage and the warning stage threshold, highlighting the potential risk if the river continues to rise. Given the steady rise of 0.5 meters per hour, it is projected to take approximately 13 hours until the red signal threshold of 19 meters is reached. This analysis underscores the importance of proactive monitoring and communication protocols within ATLANTECH disaster association emergency management framework. By staying informed and prepared for potential flooding events, ATLANTECH disaster association can effectively safeguard its operations, protect personnel, and contribute to the resilience of the surrounding community.

The red warning is not only shared with all the officers and authorities who receive the yellow warning, but it is also quickly communicated to the general public of the area using the fastest available methods. When a designated civil defense area with civil defense sirens receives a red warning, the public is alerted through the sound of the siren for three minutes. The siren emits a combination of steady and wobbling notes to convey the urgency of the situation (Moore et al., 2005).

When a red warning is issued for an area that is not under civil defense jurisdiction, the public in that area is notified through the traditional method of drum beating and the use of special messengers equipped with microphones or loudspeakers (Volz et al., 2017). Upon receiving the red warning, the administration swiftly mobilizes rescue parties and other emergency units, ensuring they are fully prepared for immediate action. When the red warning is canceled or withdrawn, the public is informed through the sound of sirens or other appropriate methods outside the civil defense area. This is done for a duration of two minutes to indicate that the danger has passed (Pappenberger et al., 2015).

6.5. ENGINEERING MEASURES FOR FLOOD FIGHTING

Learning Objectives:

- Understand emergency engineering measures for embankment protection during floods.
- Learn techniques for preventing embankment failure: sliding, overtopping, scouring.

In addition to the administrative measures mentioned earlier, several emergency construction activities are required to provide relief from the devastating impact of floods. Often, the destruction caused by floods can be attributed to the failure or breach of protective embankments. The reasons behind a bank recession and the necessary precautions have already been thoroughly examined. In this section, the emergency engineering measures that may need to be taken in the event of an exceptionally high flood that could potentially lead to significant breaches or the actual breaching of an embankment will be discussed (Lendering et al., 2016).

6.5.1. Protection Against Sliding

During flood events, it is possible for an excessive amount of water to seep through the embankment, potentially causing piping underneath the foundation. This occurrence is commonly referred to as sand boiling. Usually, this occurs at the embankment's base. It is advised to cover the boil with a ring of sand-filled bags to guard against any possible damage. The essential hydraulic gradient will be efficiently lowered by the ring's rising water level, minimizing seepage and material loss, and ultimately averting embankment failure (Yang et al., 2006).

6.5.2. Protection Against Landslide Level Slope Sliding

In order to address seepage water from a levee, it is common to incorporate slope drains along with a horizontal channel at the base of the levee. During periods of excessive flooding, the situation can quickly deteriorate, posing a significant risk of levee failure due to sliding. To prevent this, a wooden mattress is placed over the slope and loaded with gunny bags filled with sand or gravel. The heaviest load is concentrated near the berm. This measure will allow for efficient drainage and help prevent slope failure (Barla et al., 2013).

6.5.3. Protection Against Overtopping

Raising the embankment as soon as possible is essential as an emergency measure if the flood forecast shows that there is a significant possibility of the embankment being breached at a particular place. The levee will fail quickly because the floodwater will quickly erode the soil once it reaches beyond it. The embankment can be raised in a number of ways, depending on the materials that are available. An effortless solution is to utilize gunny bags loaded with sand or spoiled cement bags when the desired rise in height is minimal. Typically, box-type constructions constructed from timber or bamboo and filled with earth on top are commonly used to achieve greater heights (Morán & Toledo, 2011).

6.5.4. Protection Against Scouring of the Riverside Slope

In times of severe flooding, the existing protective measures along the riverside levee slope may prove inadequate. As a result, the slope may experience scouring, leading to a significant loss of the levee when the scouring face becomes visible above the water surface. As a precautionary measure, cribs filled with rocks or sandbags are strategically placed over the scoured areas. Implementing these measures will help reduce erosion and provide stability to the embankment (Amoros et al., 2005).

6.5.5. Breaches Closure in the Embankment

If a breach in the levee occurs, immediate action should be taken to repair it to prevent the breach from spreading and to reduce the duration of flooding in the affected areas. The restoration of the breach is conducted gradually. Initially, a ring is created around the breached area by submerging timber (ASCE et al., 2011).

SUMMARY

- The chapter titled "Flood Forecasting, Warning, and Flood Fighting" offers a thorough examination of the procedures and tactics necessary for effectively controlling and reducing the consequences of flooding. It begins by providing a comprehensive explanation of forecasting methodologies and procedures, with a particular emphasis on the crucial importance of accurate forecasts in promoting economic progress and ensuring preparedness for potential disasters.
- The analysis of runoff based on rainfall data is further examined, focusing on techniques such as rainfall-runoff relationships and unit hydrographs. These methods are used to comprehend the amount and geographical distribution of runoff generated by storm rainfall.
- The Peak Travel Time Method is emphasized in the process of predicting flood stages. This method is used to estimate the duration it takes for floodwaters to reach key areas. The practical examples provided demonstrate how forecasting methods are implemented.
- The flood warning protocols are thoroughly outlined, providing a detailed description of the signaling systems and procedures used to issue timely alerts to both departmental staff and the general public. These protocols are essential for limiting damage and aiding evacuation efforts.
- Lastly, the chapter explores engineering strategies for combating floods, which involve safeguarding against several flood-related hazards such as sliding, overtopping, scouring, and the sealing of gaps in embankments.

REVIEW QUESTIONS

1. Explain the process of hydrological modeling and its role in flood forecasting.
2. Compare and contrast the Peak Travel Time Method with other techniques used for flood forecasting.
3. How does the determination of runoff from rainfall data contribute to flood forecasting accuracy?
4. Discuss the importance of flood warning systems and their components in mitigating flood-related risks.
5. Evaluate the effectiveness of engineering measures such as protection against sliding and overtopping in flood fighting.

6. Describe the steps involved in the closure of breaches in an embankment during flood response.
7. How can historical flood data be utilized to improve future flood forecasting and response strategies?

MULTIPLE CHOICE QUESTIONS

1. Which technique involves the prediction of flood occurrence based on historical data and current meteorological conditions?

a. Hydrological modeling

b. Peak Travel Time Method

c. Forecast Reporting Example

d. Determination of Runoff from Rainfall Data

2. What does the Peak Travel Time Method primarily focus on?

a. Determining the maximum intensity of rainfall

b. Estimating the time, it takes for floodwaters to reach critical locations

c. Analyzing historical flood data

d. Calculating the volume of runoff

3. Which engineering measure is specifically designed to prevent water from overtopping embankments?

a. Protection against Sliding

b. Protection against Sliding of the Landside Level Slope

c. Protection against Overtopping

d. Protection against Scouring of the Riverside Slope

4. What is the primary purpose of flood warning systems?

a. To prevent floods from occurring

b. To accurately predict flood occurrences

c. To minimize the damage caused by floods

d. To control the flow of floodwaters

5. Which method involves estimating the amount of water runoff based on rainfall data?

a. Peak Travel Time Method

b. Forecast Reporting Example

c. Determination of Runoff from Rainfall Data

d. Protection against Sliding

6. What is the primary purpose of closure of breaches in the embankment during flood fighting?

a. To reinforce the embankment against sliding

b. To prevent water from overtopping the embankment

c. To minimize the risk of landslides

d. To restore the integrity of the embankment and control the flow of water

Answers to Multiple Questions

1. (a); 2. (b); 3. (c); 4. (c); 5. (c); 6. (d)

REFERENCES

1. Alemazkoor, N., Burris, M. W., & Danda, S. R. (2015). Using empirical data to find the best measure of travel time reliability. *Transportation Research Record, 2530*(1), 93–100.
2. Amoros, C., Elger, A., Dufour, S., Grosprêtre, L., Piégay, H., & Henry, C. (2005). Flood scouring and groundwater supply in side-channel rehabilitation of the Rhône River, France. *Arch. Hydrobiol, 155*, 147–167.
3. ASCE/EWRI Task Committee on Dam/Levee Breaching. (2011). Earthen embankment breaching. *Journal of Hydraulic Engineering, 137*(12), 1549–1564.
4. Ban, X. J., Li, Y., Skabardonis, A., & Margulici, J. D. (2010). Performance evaluation of travel-time estimation methods for real-time traffic applications. *Journal of Intelligent Transportation Systems, 14*(2), 54–67.
5. Barla, G., Antolini, F., & Barla, M. (2013). Slope stabilization in difficult conditions: The case study of a debris slide in NW Italian Alps. *Landslides, 10*, 343–355.
6. Blume, T., Zehe, E., & Bronstert, A. (2007). Rainfall-runoff response, event-based runoff coefficients, and hydrograph separation. *Hydrological Sciences Journal, 52*(5), 843–862.
7. Bodily, S. E., & Freeland, J. R. (1988). A simulation of techniques for forecasting shipments using firm orders-to-date. *Journal of the Operational Research Society, 39*(9), 833–846.
8. Borga, M., Anagnostou, E. N., Blöschl, G., & Creutin, J. D. (2011). Flash flood forecasting, warning, and risk management: The HYDRATE project. *Environmental Science & Policy, 14*(7), 834–844.
9. Brauchler, R., Liedl, R., & Dietrich, P. (2003). A travel time-based

hydraulic tomographic approach. *Water Resources Research, 39*(12), 33–88.

10. Brilly, M., & Polic, M. (2005). Public perception of flood risks, flood forecasting, and mitigation. *Natural Hazards and Earth System Sciences, 5*(3), 345–355.
11. Carsell, K. M., Pingel, N. D., & Ford, D. T. (2004). Quantifying the benefit of a flood warning system. *Natural Hazards Review, 5*(3), 131–140.
12. Cassar, G., & Gibson, B. (2008). Budgets, internal reports, and manager forecast accuracy. *Contemporary Accounting Research, 25*(3), 707–738.
13. Chang, F. J., Chiang, Y. M., & Chang, L. C. (2007). Multi-step-ahead neural networks for flood forecasting. *Hydrological Sciences Journal, 52*(1), 114–130.
14. Chatfield, C. (1978). The Holt-Winters forecasting procedure. *Journal of the Royal Statistical Society: Series C (Applied Statistics), 27*(3), 264–279.
15. Chau, K. W., Wu, C. L., & Li, Y. S. (2005). Comparison of several flood forecasting models in Yangtze River. *Journal of Hydrologic Engineering, 10*(6), 485–491.
16. Conejo, A. J., Contreras, J., Espínola, R., & Plazas, M. A. (2005). Forecasting electricity prices for a day-ahead pool-based electric energy market. *International Journal of Forecasting, 21*(3), 435–462.
17. Dazzi, S., Vacondio, R., & Mignosa, P. (2021). Flood stage forecasting using machine-learning methods: A case study on the Parma River (Italy). *Water, 13*(12), 1612.
18. Dorum, A., Yarar, A., Sevimli, M. F., & Onüçyildiz, M. (2010). Modeling the rainfall–Runoff data of Susurluk basin. *Expert Systems with Applications, 37*(9), 6587–6593.
19. Drobot, S., & Parker, D. J. (2007). Advances and challenges in flash flood warnings. *Environmental Hazards, 7*(3), 173–178.
20. Emerton, R. E., Stephens, E. M., Pappenberger, F., Pagano, T. C., Weerts, A. H., Wood, A. W., & Cloke, H. L. (2016). Continental and global scale flood forecasting systems. *Wiley Interdisciplinary Reviews: Water, 3*(3), 391–418.
21. Garen, D. C. (1992). Improved techniques in regression-based streamflow volume forecasting. *Journal of Water Resources Planning and Management, 118*(6), 654–670.
22. Han, S., & Coulibaly, P. (2017). Bayesian flood forecasting methods: A review. *Journal of Hydrology, 551*, 340–351.

23. Hao, Y., & Tian, C. (2019). A novel two-stage forecasting model based on error factor and ensemble method for multi-step wind power forecasting. *Applied Energy, 238*, 368–383.
24. Harvey, A. C. (1984). A unified view of statistical forecasting procedures. *Journal of Forecasting, 3*(3), 245–275.
25. Hawkins, R. H. (1993). Asymptotic determination of runoff curve numbers from data. *Journal of Irrigation and Drainage Engineering, 119*(2), 334–345.
26. Higatani, A., Kitazawa, T., Tanabe, J., Suga, Y., Sekhar, R., & Asakura, Y. (2009). Empirical analysis of travel time reliability measures in Hanshin expressway network. *Journal of Intelligent Transportation Systems, 13*(1), 28–38.
27. Horner, W. W., & Jens, S. W. (1942). Surface runoff determination from rainfall without using coefficients. *Transactions of the American Society of Civil Engineers, 107*(1), 1039–1075.
28. Jain, S. K., Mani, P., Jain, S. K., Prakash, P., Singh, V. P., Tullos, D., & Dimri, A. P. (2018). A brief review of flood forecasting techniques and their applications. *International Journal of River Basin Management, 16*(3), 329–344.
29. Kisi, O., Shiri, J., & Tombul, M. (2013). Modeling rainfall-runoff process using soft computing techniques. *Computers & Geosciences, 51*, 108–117.
30. Kuller, M., Schoenholzer, K., & Lienert, J. (2021). Creating effective flood warnings: A framework from a critical review. *Journal of Hydrology, 602*, 126–708.
31. Kunst, R., & Neusser, K. (1986). A forecasting comparison of some VAR techniques. *International Journal of Forecasting, 2*(4), 447–456.
32. Lendering, K. T., Jonkman, S. N., & Kok, M. (2016). Effectiveness of emergency measures for flood prevention. *Journal of Flood Risk Management, 9*(4), 320–334.
33. Mahmoud, E. (1984). Accuracy in forecasting: A survey. *Journal of Forecasting, 3*(2), 139–159.
34. Mentzer, J. T., & Kahn, K. B. (1995). Forecasting technique familiarity, satisfaction, usage, and application. *Journal of Forecasting, 14*(5), 465–476.
35. Milkovich, G. T., Annoni, A. J., & Mahoney, T. A. (1972). The use of the Delphi procedures in manpower forecasting. *Management Science, 19*(4-part-1), 381–388.
36. Moore, R. J., Bell, V. A., & Jones, D. A. (2005). Forecasting for flood

warning. *Comptes Rendus Geoscience, 337*(1–2), 203–217.

37. Morán, R., & Toledo, M. Á. (2011). Research into the protection of rockfill dams from overtopping using rockfill downstream toes. *Canadian Journal of Civil Engineering, 38*(12), 1314–1326.
38. Morss, R. E. (2010). Interactions among flood predictions, decisions, and outcomes: Synthesis of three cases. *Natural Hazards Review, 11*(3), 83–96.
39. Parker, D. J., & Handmer, J. W. (1998). The role of unofficial flood warning systems. *Journal of Contingencies and Crisis Management, 6*(1), 45–60.
40. Parker, D., & Fordham, M. (1996). An evaluation of flood forecasting, warning and response systems in the European Union. *Water Resources Management, 10*, 279–302.
41. Priest, S. J., Parker, D. J., & Tapsell, S. M. (2011). Modeling the potential damage-reducing benefits of flood warnings using European cases. *Environmental Hazards, 10*(2), 101–120.
42. Skaugen, T., & Onof, C. (2014). A rainfall-runoff model parameterized from GIS and runoff data. *Hydrological Processes, 28*(15), 4529–4542.
43. Soulis, K. X., & Valiantzas, J. D. (2012). SCS-CN parameter determination using rainfall-runoff data in heterogeneous watersheds–the two-CN system approach. *Hydrology and Earth System Sciences, 16*(3), 1001–1015.
44. Soulis, K. X., & Valiantzas, J. D. (2013). Identification of the SCS-CN parameter spatial distribution using rainfall-runoff data in heterogeneous watersheds. *Water Resources Management, 27*, 1737–1749.
45. Tariq, M. A. U. R., & Van De Giesen, N. (2012). Floods and flood management in Pakistan. *Physics and Chemistry of the Earth, Parts A/B/C, 47*, 11–20.
46. Thieken, A. H., Bubeck, P., Heidenreich, A., Von Keyserlingk, J., Dillenardt, L., & Otto, A. (2023). Performance of the flood warning system in Germany in July 2021–insights from affected residents. *Natural Hazards and Earth System Sciences, 23*(2), 973–990.
47. Volz, C., Frank, P. J., Vetsch, D. F., Hager, W. H., & Boes, R. M. (2017). Numerical embankment breach modeling including seepage flow effects. *Journal of Hydraulic Research, 55*(4), 480–490.
48. Wang, Y., Chen, Q., Sun, M., Kang, C., & Xia, Q. (2018). An ensemble forecasting method for the aggregated load with subprofiles. *IEEE Transactions on Smart Grid, 9*(4), 3906–3908.
49. Yang, Q., Senda, T., & Hirose, A. (2006). Sliding wear behavior of WC–12% Co coatings at elevated temperatures. *Surface and Coatings Technology, 200*(14–15), 4208–4212.

50. Zhang, J., Wang, Y., Sun, M., & Zhang, N. (2020). Two-stage bootstrap sampling for probabilistic load forecasting. *IEEE Transactions on Engineering Management, 69*(3), 720–728.
51. Zhao, W., Wei, Y. M., & Su, Z. (2016). One day ahead wind speed forecasting: A resampling-based approach. *Applied Energy, 178*, 886–901.
52. Zou, Y., Zhu, X., Zhang, Y., & Zeng, X. (2014). A space-time diurnal method for short-term freeway travel time prediction. *Transportation Research Part C: Emerging Technologies, 43*, 33–49.

CHAPTER 7

INTEGRATED WATER RESOURCE MANAGEMENT

LEARNING OBJECTIVES

At the end of this chapter, readers will be able to:

1. Understand the historical development and principles of Integrated Water Resource Management (IWRM).
2. Explore the key components and guiding principles of the IWRM paradigm.
3. Know about the economic dimension of water management and its implications.
4. Recognize the classification of water resources and their economic characteristics for effective management strategies.
5. Understand the concept of eco-hydrology and its role in water management.
6. Explore the dynamics of groundwater management in open-access systems.
7. Understand the ecological and hydrological effects of groundwater overexploitation.
8. Comprehend the steady-state conditions and stability in groundwater stock dynamics.

INTRODUCTORY EXAMPLE

AquaTech Solutions, a leading water technology company, operating in a region facing significant water management challenges. In recent years, rapid urbanization and industrial growth have strained local water resources, leading to concerns about water scarcity, pollution, and ecosystem degradation.

In response to these challenges, AquaTech Solutions has decided to adopt an Integrated Water Resource Management (IWRM) approach to address the complex and interconnected issues surrounding water management in the region.

By implementing IWRM principles, AquaTech Solutions aims to achieve sustainable water use, ensure equitable access to water resources, and protect the environment for future generations. Through collaboration with local communities, government agencies, and other stakeholders, AquaTech Solutions seeks to develop innovative solutions that balance the social, economic, and environmental dimensions of water management.

UNIT INTRODUCTION

Water is an essential resource that is crucial for maintaining life, upholding ecosystems, and facilitating economic activities. Integrated Water Resource Management (IWRM) is a comprehensive strategy that recognizes the mutual dependence of several facets of water management, encompassing social, economic, and ecological factors. This chapter focuses on the core principles of Integrated Water Resources Management (IWRM), examining its methodologies, perspectives, and overall structures (Mitchell, 2005).

The economic dimension of water, examining different types of environmental goods and the economics of water management will be discussed (Cardwell et al., 2006). Additionally, the concept of eco-hydrology and the importance of managing water as a public good, emphasizing the need for sustainable and equitable water management practices will be explored (Thomas & Durham, 2003).

The economic aspect of water, analyzing various forms of environmental resources, and the economics behind water management will be discussed. In addition, the field of eco-hydrology and highlighting the significance of water management as a shared resource, underscoring the necessity for sustainable and fair practices will also be analyzed (Saravanan et al., 2009).

When discussing groundwater management techniques, particular consideration is given to groundwater, an essential part of the water cycle. The groundwater management practices' hydrological and ecological implications, as well as groundwater modeling techniques and dynamic stock balancing approaches, will be explored. This chapter aims to provide insights into practical and sustainable water resource management techniques that strike a balance between the requirements of people and environmental conservation objectives through a thorough understanding of these concepts and principles (McDonnell, 2008).

7.1. INTEGRATED WATER RESOURCE MANAGEMENT

Learning Objectives:

- Understand the historical evolution of Integrated Water Resource Management (IWRM).
- Understand the key concepts and principles of the IWRM paradigm, including integration, decentralization, and sustainability.
- Gain insight into practical frameworks and tools for implementing IWRM, including the IWRM planning cycle and hydro-economic models.

7.1.1. Approaches to IWRM

The origins of the Integrated Water Resource Management (IWRM) strategy may be traced to the 1933 founding of the Tennessee Valley Authority (TVA), which combined the production of power, navigation, and flood control. Other issues like public health, recreation, and erosion management were also handled by the TVA. In 1957, the United Nations Organization (UNO) Secretary-General gave a speech on IWRM (Anderson et al., 2008). The UNO defines integration as the supporting services required to grow irrigated agriculture; nevertheless, this IWRM concept did not include the coordination of various water-related operations. The gap was filled at the 1977 Water Conference in Mar del Plata, where it was specifically stated that cooperation within the water industry was essential. However, concerns about irrigated agriculture's high water consumption and detrimental effects on the environment were not adequately addressed (Grigg, 2019).

Did You Know?

In order to guarantee sustainable water use, integrated water resource management, or IWRM, is a thorough strategy that takes into account the relationships between ecosystems, human activity, and water resources.

Traditional water management had some obvious flaws at the start of the 1990s, including problems with quality, overexploitation, environmental damage, and societal concerns. Additionally, water issues have developed into multifaceted, multisectoral, multiregional issues with several agendas, interests, and causes. In 1992, during the International Conference on Environment and Water in Dublin, four key guiding principles were established to address these problems. The "Dublin Principles" are a well-known set of ecological, institutional, gender, and economic guidelines (Allan, 2006).

The Dublin Guiding Principles played a significant role in shaping Agenda 21, a key outcome of the United Nations Conference on Environment and Development in Rio de Janeiro in 1992. The current comprehension of Integrated Water Resources Management (IWRM) with its comprehensive approach is firmly rooted in the Dublin Principles. There are multiple interpretations of IWRM, such as the widely referenced definition provided by GWP (Gallego, 2013).

"IWRM is a process that aims to foster the harmonized development and management of water, land, and associated resources to optimize economic and social well-being in a fair manner while safeguarding the long-term viability of crucial ecosystems."

It is important to understand that IWRM should not be viewed as a one-size-fits-all solution for water management. Instead, it should be seen as a flexible framework that provides a range of principles, tools, and guidelines that can be customized to suit the unique needs of a country, region, or river basin. Water resource management can be accomplished in an efficient and effective manner by tailoring IWRM to the environment (Ballweber, 2006). These are some essential ideas (Figure 7.1):

- Combine water and environmental management. Adopt a systematic strategy.
- Ensuring the active involvement of all stakeholders, including workers and the community. Focus on the societal aspects.
- Enhancing organizational capabilities.
- The presence of information and the ability to utilize it in order to predict future events.
- Full-cost pricing is supplemented by specific subsidies (Saravanan et al., 2009).

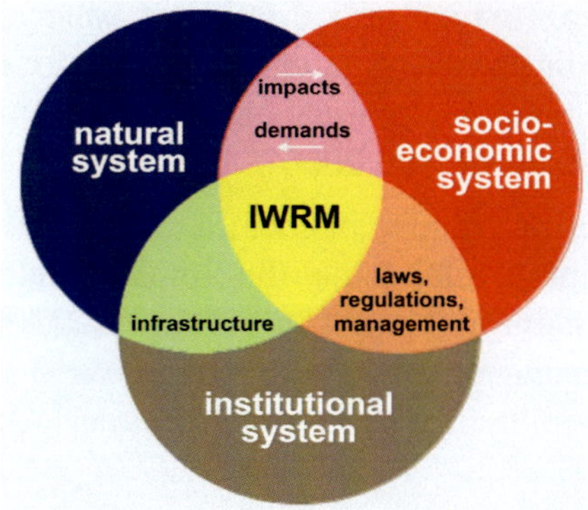

Figure 7.1. *Illustration of IWRM.*

Source: Eelco Van.; Creative Commons License.

Integrated water resources planning and management are required due to the widespread shortage, progressive degradation, and severe contamination of freshwater resources in many parts of the world, as well as the increasing encroachment of incompatible activities. Many interconnected freshwater sources, including surface water and groundwater, as well as the quantity and quality of the water must be considered in this type of integration. It is important to acknowledge both the multi-interest use of water resources and the multi-sectoral character of water resource development in the framework of socioeconomic development (Hileman et al., 2016):

- The central government's assistance through the setting up and maintaining of a suitable atmosphere;
- The adoption of the most advanced methods and technology already in use;
- Sturdy and consistent funding;
- Fair distribution of water resources;
- Water's classification as an economic good;
- Empowering women to play a bigger part in water management (Mazvimavi et al., 2008).

7.1.2. The IWRM Paradigm

The fundamental ideas of participation, decentralization, integration, and sustainability are all part of the IWRM paradigm. The comprehensive approach of the IWRM paradigm requires the coordinated management of various sectors, including agriculture, energy generation, water supply, sanitation industry, and environmental protection, which utilize or affect water resources. In addition to horizontal integration, vertical integration is also required to facilitate coordination between national, municipal, regional, and international water user groups and institutions (Allan, 2006).

Furthermore, fostering a sense of local and regional awareness is crucial for effective management, alongside the implementation of decentralized accountability and decision-making. As a result, IWRM aims to reconcile top-down and bottom-up management. Additionally, IWRM seeks to support water user associations and community-based organizations (Allouche, 2016) (Figure 7.2).

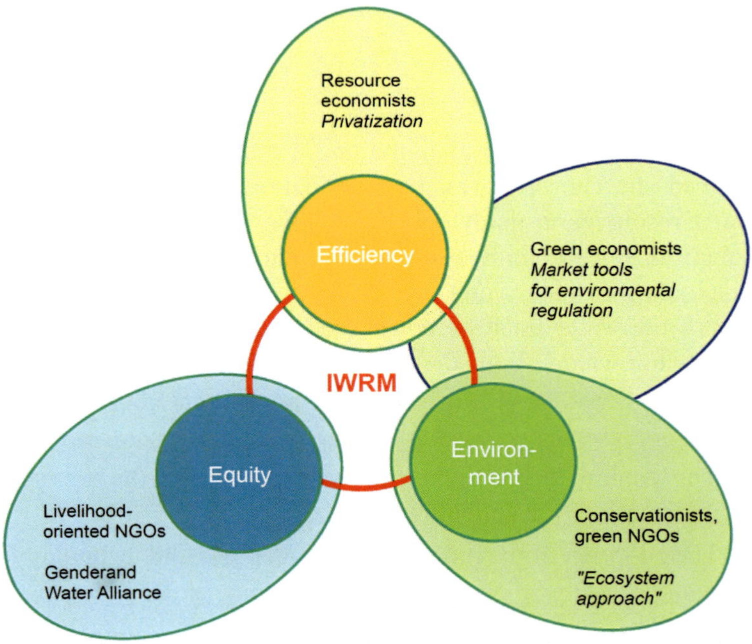

Figure 7.2. *Illustration of IWRM and its three 'E.'*

Source: Francois Molle; Creative Commons License.

In IWRM, sustainability plays a vital role that extends beyond mere preservation of the natural system. It also covers financial and economic aspects. As an example, it is necessary to make resource allocation decisions based on the economic value of water. Consequently, it is essential to accurately determine the true cost of water (Del Moral et al., 2014).

Equity, ecological integrity, and efficiency are the three key policy goals of IWRM, commonly referred to as the three 'E's:

1. **Equity:** Access to water of sufficient quantity and quality is a fundamental right that should be available to all individuals (Tremblay, 2011).
2. **Ecological Integrity:** Adequate amounts of water with satisfactory quality should endure in the environment. Water should be utilized in a sustainable manner to ensure that future generations can access it in a manner similar to the present age.
3. **Efficiency:** Water is limited and delicate, thus it needs to be used as efficiently as possible. The water service should achieve cost recovery. Pricing for water needs to reflect its economic worth (Pahl-Wostl et al., 2011).

The Global Water Partnership (GWP) has developed a toolkit with three primary categories to facilitate the implementation of IWRM principles in practice: institutional roles, management instruments, and enabling environment:

- An "enabling environment" is one that safeguards the property and rights of all parties involved as well as public resources. The broad framework of national laws, regulations, and policies falls under this category (Lautze et al., 2011).
- The institutional roles involve taking into account a wide range of formal laws and regulations, traditions and practices, concepts and data, and networks of interest and community groups. These elements collectively provide the institutional framework or environment that decision-makers work within (Grigg, 2019).
- The operational tools for monitoring, regulating, and assisting decision-makers are part of the management tools.

7.1.3. IWRM General Framework

To implement the IWRM paradigm effectively, it is recommended to follow an IWRM planning cycle. Overall, the complexity of the water cycle and the interconnectedness between the water sector and other sectors necessitate the use of specialized approaches to incorporate social, environmental, and economic considerations at the watershed level (Cardwell et al., 2006). The IWRM paradigm equips us with a range of interdisciplinary tools derived from various fields such as natural water science (including geohydrology, hydrology, and meteorology), engineering, and social sciences including sociology, political science, and economics (Grigg & Grigg, 2016).

Many of these methods, like decision-supportive systems or optimization models, etc., rely on mathematical models to capture the complexity involved. Hydro economic models simulate the geographic distribution of different water system elements using node-link networks and simulation or optimization models, which are employed in integrated water resource management (Kalantari et al., 2018). These elements include natural water bodies like seas, lakes, aquifers, and river sections, as well as artificial water bodies such as canals. Infrastructure like dams, pipelines, wells, pumps, and purification plants are also considered, along with human and artificial impacts on the water system, such as point and non-point pollution sources (Mehta & Movik, 2014) (Figure 7.3).

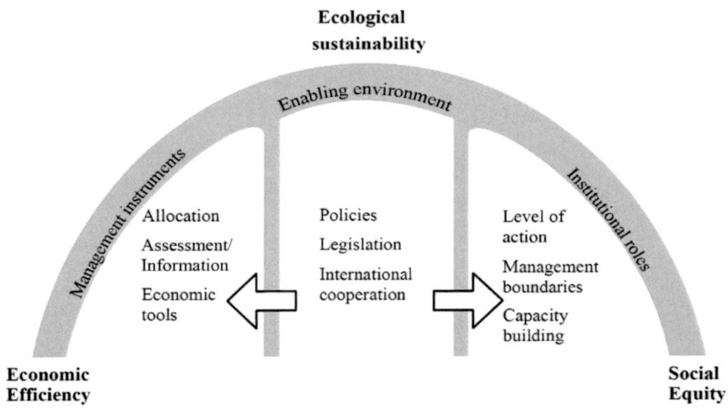

Figure 7.3. *Illustration of the general framework of IWRM.*

Source: Andreas, H.; Creative Commons License.

The subsequent sections present comprehensive yet uncomplicated models that systematically address the primary issues of Integrated Water Resources Management (IWRM) (Al-Radif, 1999). The IWRM framework covers specific subjects and economic instruments, including pricing policies and the management of transboundary rivers (Ballweber, 2006).

7.2. ECONOMIC DIMENSION OF WATER

Learning Objectives:

- Understand the economic valuation of water and its implications.
- Know the classification of water as an environmental good.
- Understand the economic dimensions of water use and management.

During periods of water scarcity, the economic aspect of water becomes increasingly important. This has been mentioned in the Dublin declaration that was previously cited. According to the principle, water is valued for its economic potential in various applications and should be treated as a commercial commodity. Treating water as a valuable commodity is a crucial approach to ensure optimal and fair utilization, as well as to promote the preservation and safeguarding of water resources (Siwar & Ahmed, 2014).

Highlighting the economic value of water does not suggest that it is only meant for private use. Furthermore, it does not imply that the management of water supply should be handed over to private companies (Cohen & Sullivan, 2010). It suggests that the water cycle should be treated as a limited resource that needs to be carefully managed. One notable distinction of this scarcity from other scarce goods, such as precious old paintings or fossil fuels, is that it arises from a deliberate political choice to avoid overexploiting the water cycle (Isengard, 2008).

The recognition of limited resources stems from the commitment to maintaining sustainability. This statement emphasizes the importance of implementing necessary management steps to ensure sustainability and economic efficiency in an institutional setting. In the following subsection, the different aspects of the water cycle from an economic standpoint will be explored (Lundqvist et al., 2003).

7.2.1. Environmental Goods Types

Water may be owned by society. There are numerous instances globally of watersheds that are owned and managed collectively. Reports from irrigation cooperatives in the Philippines and Spain, for instance, provide insight into how irrigation water is distributed within a communal institutional framework that includes monitoring and dispute-resolution procedures. Furthermore, there are market-based solutions, such as water market institutions in Australia and the Southwest of the United States of America (Liu et al., 2022).

Water ownership in that region is commonly held privately according to established property rights and is exchanged through spot and forward markets. Hence, it is crucial to differentiate between water as a commodity with economic value and water as an individual-owned resource. A private good is characterized by its inherent rivalry and the capacity to exclude others through the enforcement of property rights (Stenger et al., 2009).

Consider a scenario where farmer A irrigates his fields, leaving farmer B with a shortage of water (rivalness). However, for farmer A to access and use water, it is necessary for him to have exclusive rights to it, which means other users are excluded. Nevertheless, water is not solely considered a private commodity. Undoubtedly, the water cycle plays a crucial role in sustaining the well-being of individuals living in a watershed through its fulfillment of various essential ecosystem functions. Environmental economists study how the water cycle and other natural resources, such as soil, nutrients, and vegetation, interact to create the ecosystem in a watershed. These ecosystem services contribute to societal benefits by generating economic value (Hervani et al., 2017).

For example, a wetland serves the dual purpose of reducing flood damage and providing a space for recreational activities. Forests play a crucial role in replenishing groundwater and positively impacting the local climate through evapotranspiration. From an economic perspective, these essential functions have a significant impact on society and the economy, going beyond the mere practical use of water for irrigation or household consumption (Livesey, 2010). Several of the water ecosystem services mentioned above are considered public goods. Public goods are defined by their unique characteristics of non-rivalry and non-excludability. The local water cycle, for example, plays a crucial role in maintaining the stability of the watershed's microclimate, which is essential for the livelihood of the residents. Everyone in the local population benefits from this positive ecosystem function, and no one is left out (Dlamini, 2012).

On the other hand, let's consider the case of how a forest habitat or wetland can help protect against floods. Once again, all neighbors benefit from the shelter provided. The division of natural resources and their services into several categories of economic products is shown in Table 7.1. Each form of management strategy needs to be customized to guarantee a productive and ecologically sound supply for the community (Balineau & De Melo, 2013).

Table 7.1. *Types of Environmental Goods*

	Rival	**Non-Rival**
Excludable	Oil, gas, timber, private good food,	National Park, golf club lane, club goods swimming pool.

Non-Excludable	Deep-sea fishing, open-access resources, and ecosystem services	Carbon-absorbing potential of the rainforest for the general good and ecosystem services.

Source: Wyne C Hubber; Creative Commons License.

Consider natural resources as private goods. Resources such as oil, timber, and gas are commonly traded in markets. Yes, they are considered private goods because they have clear property rights and are subject to competition. When it comes to club goods, the situation is quite similar. You have to pay for the services they offer, but unlike private goods, your use of the goods does not limit the availability for other club members. Now let's consider the situation with open-access resources. Deep-sea fishery serves as an excellent example (De Alwis, 2014).

It is not possible to stop someone from fishing outside of the 200 nautical mile exclusive economic zone from the terrestrial baseline. As a result, their fishing is competitive but there is no excludability. Eco-services and natural resources may also be considered public goods since they benefit everyone and cannot be withheld from anybody, even if they are not paid for. The ability of the rainforest to absorb carbon is a prime example (Kerkhof et al., 2010).

What is the significance of this classification for economists? To explain the significance of these differences, consider the case of deep-sea fishing. The absence of exclusivity in fishing grounds results in the excessive exploitation of fish populations. There is a concerning issue with the excessive number of trawlers that are currently operating without considering the impact of their fishing activities on the fish population. Furthermore, if a fisherman is worried about the long-term sustainability of fish stocks, reducing their fishing efforts not only harms their interests but also fails to support the conservation of the fish population (Whitehead & Blomquist, 1991).

His colleagues will catch the fish that he chooses not to fish. Clearly, a free market system without regulations is not an appropriate model for effectively and sustainably managing fisheries. Instead, it necessitates public intervention (De Groot et al., 2002). Likewise, the provision of public goods should not be entrusted to a market. Typically, when public goods are provided through a market with uniform pricing for all customers, there is a tendency for these things to be inadequately provided. Since every consumer must be able to consume the commodity, the provider cannot generate adequate revenues. Once more, the implementation of a free-market approach would lead to an unsatisfactory outcome, which is occasionally referred to as market failure (Merlo et al., 2000).

Did You Know?

Key elements of the economic dimension of water include seeing water as an asset and managing it equitably and effectively to encourage resource conservation.

7.2.2. Economic of Water

The classification shown in Table 7.2 helps to further define the economic aspect of water use. The different ways that water is used demonstrate the water cycle's economic aspects. Various institutional frameworks are necessary to ensure adequate water services that meet the unique needs arising from different water uses and diverse industrial patterns (Barbier, 2004).

Table 7.2. *Economic Dimensions of Water*

	Rival	**Non-Rival**
Excludable	Private good. Providing clean and safe water for households, supporting farmers with irrigation, and managing groundwater extraction in a controlled manner.	Club goods. There are different types of in-stream uses available, while other recreational activities are limited to club members only.
Non-Excludable	Open-access resources. Groundwater extraction and the use of rivers as wastewater sinks.	Public good. Enhancing microclimate stability, managing soil conditions, promoting nutrient retention, fostering diverse habitats and biodiversity, and utilizing wetlands for flood control.

Source: Baxter, E.; Creative Commons License.

It is important to clarify that the concept of a private good is not associated with the ownership of a water resource by an individual. It solely relates to its attributes of being capable of excluding others from its use and being subject to rivalry. The regulations related to the supply of a particular volume of water have not yet been defined. Perhaps she had acquired that water from a private vendor or obtained it at no cost from a government entity (Rogers et al., 2002).

On the other hand, let's consider the example of groundwater extraction. The groundwater may be subject to common property law, which designates it as a communal resource owned by the municipality or entire community. Water extracted from a groundwater reservoir is distributed to the members based on established rules, ensuring it is treated as a private good. Water might be obtained for free up to a specific amount, establishing a system of rationing, or it could be accompanied by payments dependent on the amount collected. For this situation, it is crucial that local institutions, such as a municipality or a water cooperative, ensure the funding for the essential technical infrastructure, including pipes, pumps, and energy (Savenije, 2002).

Integrated Water Resources Management (IWRM) is a holistic approach that takes into account the intricate relationships between water resources, ecosystems, and human activities to promote the long-term viability of water usage (Dolan et al., 2021).

Practice Problem

In a region grappling with water scarcity, devise a pricing strategy to allocate water among agriculture, industry, and households while maximizing economic efficiency and ensuring equitable access. Determine water demand elasticity, calculate marginal extraction costs, and incorporate environmental externalities to establish a tiered pricing structure that discourages wasteful usage, promotes conservation, and reflects the true economic value of water.

Solutions to Practice Problem

Conduct surveys to gauge water demand elasticity across sectors, accounting for factors like energy prices and infrastructure maintenance. Analyze the costs of water extraction from surface and groundwater sources, integrating environmental costs to determine true extraction costs. Implement a tiered pricing model that incentivizes conservation and efficiency, offering subsidies for essential needs. Monitor usage patterns to adjust pricing policies and optimize economic, social, and environmental outcomes, ultimately ensuring sustainable water management.

7.3. WATER MANAGEMENT AS A PUBLIC GOOD AND ECO-HYDROLOGY

Learning Objectives:

- Understand the concept of eco-hydrology and its importance in water management.
- Understand the economic implications of managing water as a public good.
- Comprehend the optimization models for allocating water between private consumption and ecosystem services.

The water cycle is essential for delivering environmental services besides supplying water for human use and as a raw material for industrial. For example, it produces a vapor that helps maintain the microclimate and supports the natural system of water and trees. Forests generate ecosystem services such as flood regulation, water purification, and the provision of habitats for diverse species. Conversely, forests require water and land resources (Fritz, 2011).

An integral part of the overall ecosystem is the water cycle. There are several interrelated natural cycles in this ecological system.

In addition to the hydrological cycle, there is also the carbon-oxygen cycle, which includes two phases: the photosynthesis phase, in which carbon dioxide is transformed into oxygen, and the decomposition phase, in which organic molecules are broken down into water and carbon dioxide. Nonetheless, the nitrogen cycle is necessary for the growth and decay of plants (Jr Wescoat, 2013).

These cycles make up the nutrient cycles that many living systems are a part of on an ecological level. From a broader perspective of Integrated Water Resources Management (IWRM), these cycles are influenced by the configuration of the local hydrological cycle and the organization of land use. Consequently, certain scientists argue that a complete policy strategy must encompass these connections (Zalewski, 2015).

A management system that follows a comprehensive integrated approach is known as eco hydrology; a concept introduced by Rodriguez. The entire climate-soil-vegetation system is encompassed by it. Therefore, it is crucial to establish a strong connection between landscape planning and water resource management. From this perspective, IWRM extends well beyond the effective supply of water for personal use by individuals or businesses (Meran et al., 2021). Efficient river management involves not only coordinating water abstraction but also ensuring the availability of water for local ecosystem services. By guaranteeing the sustainability of the many natural cycles already outlined, these services are preserved. In addition, these services offer prominent benefits including leisure activities like hiking, camping, fishing, and just spending time in nature, which enhances the cultural landscape both acoustically and visually (Wagner & Breil, 2013).

From this perspective, it is important to recognize that Integrated Water Resources Management (IWRM) encompasses more than just the management of water flows for personal purposes. Considering the economic perspective, water management acknowledges the ecological services offered by water, acknowledging its value as both a private and public asset. As an example, the water cycle plays a crucial role in stabilizing the microclimate, benefiting not only the inhabitants of a watershed but also those beyond its boundaries (Schiemer, 2016).

Let's explore two different options for water management. Water can be either extracted for personal use or conserved to support the ecosystem. For a more streamlined analysis, one can assume that the measurement of ecosystem services can be done using a utility function, i =, and this function is influenced by these services. The variable "i" represents the index of an individual, whereas the variable "n" represents the population size of the watershed.

Ecosystem services encompass the intricate interconnectedness of the hydrological cycle, flora, and geological composition within a given watershed. The efficacy of E is not just determined by the policy instruments of Integrated Water Resources Management (IWRM) but is also influenced by other economic factors that affect the environment and shape the landscape, such as soil sealing and agricultural vegetation (Savenije et al., 2014).

The following discussion will center on the distribution of water for private purposes and public services. Recognize the evapotranspiration of vegetation, commonly referred to as green water, as a valuable resource that benefits all members of a community equally through the vapor it produces. Therefore, the usefulness of ecosystem services is determined by the evapotranspiration, ET, as shown in Eqn. (1).

$$U_i = U_i(ET) = U_i(\gamma_1 S), \quad i = \{1, 2, \ldots, n\}, \text{ and } U'' < 0 \tag{1}$$

The private water consumption advantages are reflected in a benefit function, B_i, that quantifies the profits or benefits derived from water consumption (Nuttle, 2002).

$$B_i = B_i(w_i), \quad i = \{1, 2, \ldots, n\} \tag{2}$$

Linear dynamic mass equations can effectively capture the hydrology.

$$\frac{dS(t)}{dt} = R + P - \gamma_1 S - \gamma_2 S - \sum_{i=1}^{n} w_i \tag{3}$$

Limiting ourselves to a steady-state analysis, or assuming that the local hydrological cycle is an equilibrium where dS(t)/dt = 0, will simplify the optimization process. Finding S, obtain:

$$S = \frac{R + P - \sum_{i=1}^{n} w_i}{\gamma_1 + \gamma_2} \tag{4}$$

The equation demonstrates that in order to solve the IWRM problem, the social welfare function must be defined and is given in Eqn. (5). *S* is dependent on the water allocation to consumers, w_i with, i = [1, 2, …, n] (Figure 7.4).

$$SWF = \sum_{i=1}^{n} [B_i(w_i) + U_i(\gamma_1 S)] \tag{5}$$

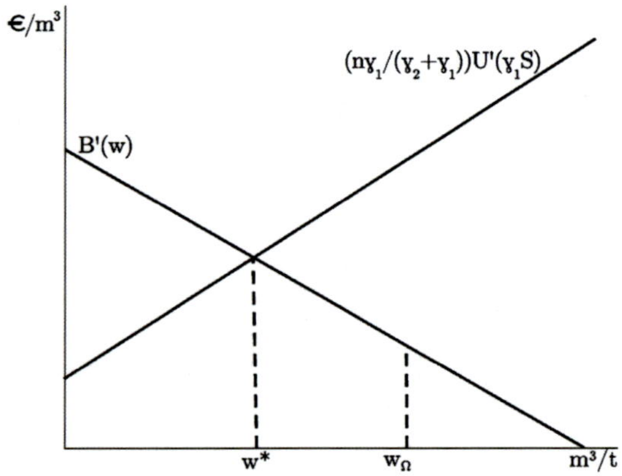

Figure 7.4. *Efficient distribution of water as a valuable resource.*

Source: Own illustration; Ramesh, S. V.; Creative Commons License.

One of the main objectives is to optimize the social welfare function by considering the impact of private consumption on the ecosystem's water resources (refer to Eqn. (4)). Deriving the optimal conditions for everyone is a relatively simple task (Ganoulis, 2021).

$$B_i'(w_i) = \frac{\gamma_1 \sum_{i=1}^{n} U_i'(\gamma_1 S)}{\gamma_1 + \gamma_2}, \quad w_i, i = \{1, 2, \ldots, n\} \tag{6}$$

The Samuelson rule, which outlines how to calculate the ideal quantity of a public good, is reflected in this set of equations. The primary argument is that since each liter of private water usage is linked to costs resulting from the marginal loss of ecosystem services for every resident, the private marginal benefit should equal the total of the marginal benefits of overall residents. It is therefore essential that both values be optimally balanced. The collection of equations can be reduced to a single figure by assuming that all inhabitants have the same valuation functions (Eberbach, 2003). Notice that 7.6 in the equation above reduces to:

$$B'(w) = \frac{\gamma_1 n U'(\gamma_1 S)}{\gamma_1 + \gamma_2} \tag{7}$$

Where Eqn. (4) defines S (Figure 7.2). The intersection of both marginal valuation curves is the ideal water consumption. The optimal evapotranspiration, or ET, is obtained by inserting $w*$ into Eqn. (4). ET interacts with all the other natural cycles that were previously discussed. Consequently, the establishment of a microclimate contributes to the maintenance of environmental services and the welfare of the local human population. The relationship between the marginal valuation curves might change over time; for example, population size and cultural characteristics can cause to shift and migrate to the right (Shang et al., 2024).

Water management has the option to choose from a specific range of (ET, w) combinations. Landscapes can be molded in diverse ways, influenced by cultural traditions and the biological requirements of society. Nevertheless, there are limitations. Outside of these limits, there will be permanent alterations to the environment (Chen et al., 2023).

The local ecological system may become unfit for human life, like a desert, if it surpasses specific ecological tipping points. The boundary is shown in Figure 7.2 as $w\Omega$. When the amount of water taken from a source exceeds a certain threshold, the process of evapotranspiration diminishes significantly, leading to a total transformation of the microclimate. The ecological system transforms into a semiarid or arid zone, resulting in negative repercussions for society (Tague et al., 2019).

7.4. GROUNDWATER MANAGEMENT

Learning Objectives:

- Understand the challenges of groundwater management in open-access systems.
- Explore the dynamics and stability of groundwater stocks in aquifers.
- Analyze the hydrological and ecological effects of overexploitation in groundwater economies.

7.4.1. Groundwater Model

Groundwater plays a crucial role in meeting water needs in various parts of the globe. In areas where surface water resources are limited due to the lack of lakes and rivers, groundwater becomes the sole source of water. In areas where water is limited, groundwater is frequently utilized for agricultural needs, particularly irrigation (Asher et al., 2015).

Groundwater extraction is a significant concern, particularly in regions with limited water resources. This issue can occur because of insufficient institutions, non-enforceable water rights, legal restrictions, or high transaction costs for assessing aquifer regulation. This issue is characterized by:

- Ensuring universal access to the aquifer: There is a lack of regulation for accessing groundwater resources. Consequently, any individual can withdraw groundwater from the aquifer.
- Competing for water resources: The amount of water extracted by one person cannot be used by another (Scibek et al., 2007).

In this institutional setting, there is a concern about the potential overexploitation of groundwater resources due to unrestricted access. This issue is of great significance in various regions across the globe, particularly in areas where groundwater serves as the primary water source. These regions are characterized by limited water availability per person, arid meteorological conditions with low precipitation rates and high potential evaporation, and significant water consumption by sectors such as agriculture (de Graaf et al., 2015).

The open-access characteristics described here create incentives for the excessive use of groundwater resources, as shown by the algebraic model presented below. As shown in Figure 7.5, an aquifer with a limited supply of groundwater, referred to as S, is assumed. This stock is supplied by a specific natural inflow, referred to as R. In addition, there is a specific quantity of water, denoted as r, which exits the groundwater supply because of natural flow processes. For clarity, it is presumed that these inherent movements remain consistent throughout a period (Condon et al., 2021).

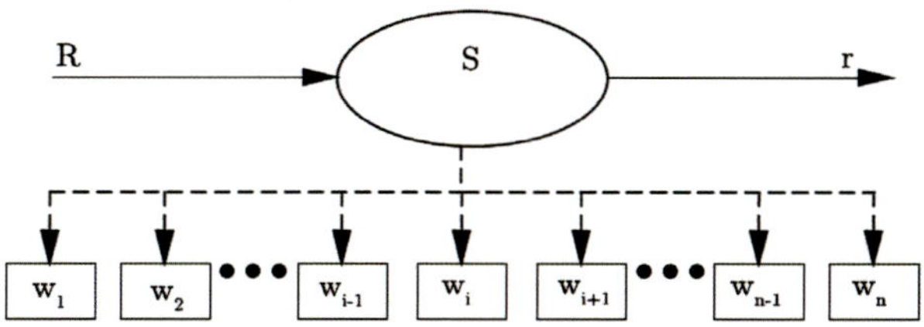

Figure 7.5. *Illustration of simple groundwater model scheme.*

Source: Sataro Oishi; Creative Commons License.

Did You Know?

Groundwater management involves complex modeling and dynamic stock balance to prevent overexploitation and minimize ecological damage, highlighting the delicate balance between human needs and environmental preservation.

Many water extractors regularly use the aquifer, either to supply water to themselves or to sell to other water consumers. Whether an extractor distributes or uses the water directly for their personal needs is not relevant to the investigation. The variable w_i, where *i* represents the quantity of groundwater collected from the aquifer by a single groundwater extractor. The total quantity of water extracted by all users during a specific duration is equivalent to the combined volume of water taken from the aquifer (Moore & Doherty, 2006).

$$W = \sum_{i=1}^{n} w_i \tag{8}$$

Assuming that all extractors have the same qualities, the equation above becomes simpler.

$$W = n \cdot w_i \tag{9}$$

The Eqn. (10) represents the demand function for all consumers in the groundwater basin, using the parameters and b. Parameter a set the choke price, while parameter b controls the gradient of the demand function (Maxwell & Miller, 2005).

$$P = a - b \cdot W \tag{10}$$

The process of extracting water quantities from the aquifer, w_i, using extractor i incurs expenses that can be characterized by the following cost function:

$$C(w_i, S) = (c - \sigma \cdot S) \cdot w_i \tag{11}$$

The amount of water extracted, , and the quantity of the aquifer's groundwater reserve, S, both affect the extraction costs. This is because a reduced water reserve causes the groundwater table to drop, necessitating higher pumping heights. Higher costs are consequently incurred for pumping.

The impact of a water stock's size on extraction cost is influenced by a level parameter called σ. Higher levels of σ result in greater sensitivities of extraction cost to the water stock S. In addition, there is a cost function parameter, denoted as c, which represents the theoretical cost rate in the event of complete depletion of groundwater resources in the aquifer (Arnold et al., 1993).

Given the similarities between the characteristics of groundwater resources and open-access goods, it is reasonable to assume that water extraction would not yield any profit. This assumption appears to be quite reasonable: Every groundwater extractor who is competing with others for scarce and unregulated groundwater resources has a chance to enter the market or raise their extraction quantities if there is still room for profit (Gaur et al., 2011).

As the amount of water being extracted continues to rise $(W \uparrow)$, it is expected that the market price for groundwater will decrease $(p \downarrow)$. Therefore, the value of using extracted groundwater diminishes as more is extracted. In addition, there is a noticeable rise in extraction costs $(C \uparrow)$, due to the increasing difficulty of extracting groundwater from a declining groundwater table in the aquifer (Brunner et al., 2007).

As a result, the supplier's profits are negatively impacted when extraction amounts increase. This process will continue until there are no longer any profits to be made in the market. Therefore, based on the assumptions mentioned earlier, the price for water will be equal to the average extraction cost, as shown in Eqns. (12) and (13).

$$P = \frac{C(w_i, S)}{w_i} \tag{12}$$

$$\Rightarrow \; a - b \cdot W = c - \sigma \cdot S \tag{13}$$

Eqn. (14) can be used to calculate the total amount of water extracted from an aquifer using the method previously described:

$$W = \frac{a - c + \sigma \cdot S}{b} \quad (14)$$

Practice Problem

In a region heavily reliant on groundwater for agricultural irrigation, the over-extraction of groundwater has led to a decline in the water table, resulting in adverse hydrological and ecological effects such as the drying up of wetlands and streams, and the depletion of groundwater-dependent ecosystems. Propose a management strategy that integrates hydrological and ecological considerations to address these issues effectively.

Solutions to Practice Problem

To address the hydrological and ecological challenges posed by over-extraction of groundwater, an integrated water resource management approach is essential. This strategy involves conducting a comprehensive hydrological assessment to understand groundwater dynamics, implementing sustainable groundwater use practices including monitoring and regulation, prioritizing ecosystem restoration efforts focusing on sensitive areas, promoting water conservation measures to reduce demand, engaging stakeholders in decision-making processes, and establishing robust monitoring and adaptive management systems to evaluate and adjust management strategies as needed. By integrating hydrological and ecological considerations, this approach ensures sustainable groundwater management while preserving essential ecosystems and water resources for future generations.

7.4.2. Dynamic Stock Balance for Groundwater

The hydrological cycle, which requires that all water volumes must be in equilibrium, allows for the establishment of a dynamic stock balance for the groundwater supply. This suggests that the groundwater table will increase in response to inflows into the groundwater stock that are greater than the amount of water outflowing from the groundwater stock, and vice versa (Dhungel & Fiedler, 2016).

As a result, the fluctuation in the groundwater stock *S(t)* is determined by the difference between the water volumes entering and leaving the aquifer. Given the correlation, it is important to note that the groundwater stock in the aquifer is subject to fluctuations over time. Therefore, it is necessary to adjust the parameter to be time-dependent, denoted as *S(t)*. As a result, the quantity of water extracted can vary over time, as shown in Eqn. (14), and should be denoted as *W(t)* (Huang et al., 2022).

The only factor that determines the inflow into the groundwater stock is the natural inflow, or R, which is taken to remain constant across time for the sake of simplicity. There are two components to the groundwater stock outflow: It is influenced by two factors: the total amount of water extracted by people, *W(t)*, and the natural outflow, r, which results from flow mechanisms and is thought to remain constant throughout time. The following dynamic stock balance is the outcome of this.

$$\dot{S}(t) = (R - r) - W(t) \tag{15}$$

$S^{\cdot}(t)$ is the algebraic formulation of the dynamic stock balance obtained by plugging Eqn. (14) into Eqn. (15).

$$\dot{S}(t) = (R - r) - \frac{a - c + \sigma \cdot S(t)}{b} \tag{16}$$

In the steady state, the inflows and outflows of groundwater are perfectly balanced, i.e., $W(t) = R - r$. This means that there are no changes in the groundwater stock over time, i.e., $\dot{S}^{\cdot}(t) = 0$. It is also referred to as the steady-state condition. According to this assumption and the Eqn. (16), one can determine the size of the water stock, *S**, that meets the steady-state condition, $\dot{S}^{\cdot}(t) = 0$. This can be found by referring to Eqn. (18).

$$\dot{S}(t) = (R - r) - \frac{a - c + \sigma \cdot S(t)}{b}$$

$$\Rightarrow \quad 0 = (R - r) - \frac{a - c + \sigma \cdot S(t)}{b} \tag{17}$$

$$\Rightarrow \quad S^* = \frac{b \cdot (R - r) - (a - c)}{\sigma} \tag{18}$$

Formally showing the stability of the steady-state situation can provide a more comprehensive understanding of the mechanism. For a stable steady state to exist, the groundwater stock, *S(t)*, must ultimately approach and stabilize at the steady-state stock, S*, over time, regardless of any initial divergence from the steady state, *S(t)* – S*. Therefore, Eqn. (17) is restated, leading to Eqn. (19) (Asadabadi et al., 2020) (Figure 7.6).

$$\dot{S}(t) = (R - r) - \frac{a - c + \sigma \cdot S(t)}{b} \tag{17}$$

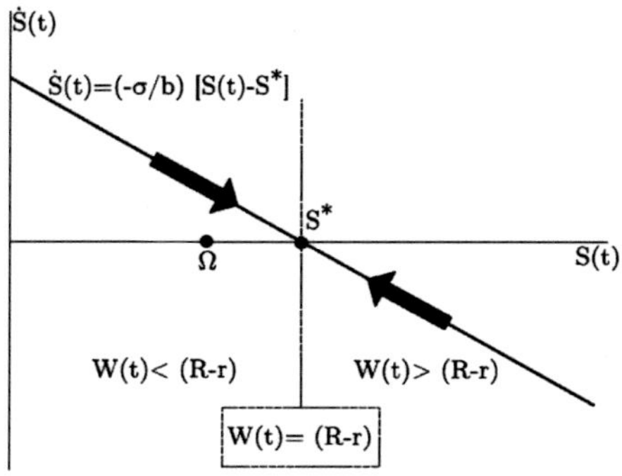

Figure 7.6. *Illustration of dynamic stock balance phase diagram.*

Source: Philip, B.; Creative Commons License.

$$\Rightarrow \quad \dot{S}(t) = \left(-\frac{\sigma}{b}\right) \cdot \left[\frac{a}{\sigma} - \frac{c}{\sigma} + S(t) - \frac{b}{\sigma} \cdot (R - r)\right]$$

$$\Rightarrow \quad \dot{S}(t) = \left(-\frac{\sigma}{b}\right) \cdot \left[S(t) + \frac{(a - c) - b \cdot (R - r)}{\sigma}\right] \tag{18}$$

By substituting Eqn. (17) into Eqn. (18), one can obtain a functional expression for the dynamic stock balance (Roumasset & Wada, 2013). This equation describes how the rate of change of the groundwater stock, $S^{\cdot}(t)$, is influenced by deviations from the steady-state groundwater stock, $(S(t) - S*)$.

$$\Rightarrow \quad \dot{S}(t) = \left(-\frac{\sigma}{b}\right) \cdot [S(t) - S^*] \tag{19}$$

According to Eqn. (19), it can be demonstrated that the steady state serves as a stable point towards which the groundwater stock gradually converges (Merrill & Guilfoos, 2018). This happens when the groundwater stock's temporal change, $S^{\cdot}(t) > 0$, turns positive when it falls below the steady-state stock, as shown by $(S(t)\ S*) < 0$. Consequently, over time, the groundwater supply would progressively rise (Hellegers et al., 2001).

On the other hand, if the groundwater stock exceeds the steady-state stock, i.e., $(S(t) - S*) > 0$, meaning there is more water than usual, there will be a gradual decline in the groundwater stock. In this scenario, the negative temporal change in groundwater stock $S^{\cdot}(t) < 0$ is the cause of this (Khan et al., 2009).

According to the stability study, the groundwater stock, *S(t)*, will eventually reach a stable condition, S∗, which is usually referred to as the long-term groundwater stock. The extent to which the current stock deviates from its stable state, as indicated by $|(S(t) - S*)|$directly influences the rate at which the groundwater stock, |S(t)|, changes. This, in turn, results in a quicker return to the steady-state stock $S*$. Figure 7.4 shows the phase diagram of the dynamic stock. Eqn. (20) is utilized as a starting point for determining the steady-state stock (Kalf & Woolley, 2005).

$$S^* = \frac{b \cdot (R - r) - (a - c)}{\sigma} \tag{20}$$

7.4.3. Hydrological and Ecologic Effects

"After having studied the dynamic properties of the model, we would like to study the hydrological and ecologic effects of the open- access groundwater economy. To this end, we introduce a threshold value Ω. This threshold value represents the height of the critical groundwater level. This results in two cases:

1. When the condition $S^* \geq \Omega \rightarrow b\,(R - r) - (a - c) > \sigma\,\Omega$, is met, it ensures that the long-term water stock remains above the threshold value. There is no overexploitation happening in the aquifer being discussed. The stability of the microclimate, vegetation, and other hydrological functions is maintained (Li et al., 2015). It is important to note that this situation is not a result of a water management approach focused on sustainability goals, but rather due to the high costs associated with pumping. Low-productivity pumping technology is used to safeguard the water cycle and the environment (Hayashi & Rosenberry, 2002).
2. Nevertheless, if $S^* < \Omega \rightarrow b\,(R - r) - (a - c) < \sigma\,\Omega$, the water usage is not properly managed and regulated, the consequences could be devastating for the local ecology. Over time, the occurrence of overdrafts will lead to negative impacts on the soil quality, regional climate, and the local hydrological cycle. These impacts can have long-term repercussions. The idea of hysteresis is frequently explored in ecology. It might be too late to completely restore the environment to its original state, even with the greatest of intentions to lessen the harm in the future (Rolls et al., 2012).

In this subsection, the analysis conducted using models enables us to draw conclusions about the issues of scarcity that may arise when an aquifer is overexploited without proper regulation or limitations. The risk of overexploitation is not solely dependent on the cost rate parameter, σ, which represents the effect of the water stock size on extraction costs. Instead, potential overexploitation is influenced by various other parameters (Rosenberg et al., 2000).

SUMMARY

- The chapter delves into a thorough examination of Integrated Water Resources Management (IWRM), encompassing its core principles, economic factors, and ecological considerations. It explores different approaches to Integrated Water Resources Management (IWRM), explaining its concept and offering a broad framework for its application.
- The chapter explores the economic aspect of water, providing a comprehensive analysis of various environmental goods and the complex economics surrounding water. It delves into topics such as valuation and allocation mechanisms.
- The importance of eco-hydrology in water management as a public good is emphasized, recognizing the connection between ecological processes and human activities.
- Detailed discussions on groundwater management strategies include modeling techniques, dynamic stock balance methodologies, and considerations of hydrological and ecological impacts.
- The chapter highlights the comprehensive nature of Integrated Water Resources Management (IWRM) and the significance of integrated approaches for ensuring sustainable water resource management.

REVIEW QUESTIONS

1. What are the principles of Integrated Water Resources Management (IWRM)?
2. How do the approaches to IWRM differ from each other?
3. What is the significance of the IWRM paradigm in water management?
4. What are the components of a general framework for IWRM?
5. Explain the concept of environmental goods and their types, with a focus on water.
6. How do economic factors influence water management decisions?
7. What is eco-hydrology, and how does it relate to the management of water as a public good?
8. What are the challenges associated with groundwater management, and how can they be addressed?

MULTIPLE CHOICE QUESTIONS

1. **What does IWRM stand for?**
   a. Integrated Water Resources Management
   b. International Water Resource Management
   c. Individual Water Resource Management
   d. Integrated Water Reservoir Management
2. **Which approach to IWRM emphasizes the interconnectedness of water systems?**
   a. Fragmented approach
   b. Sectoral approach
   c. Integrated approach
   d. Isolated approach
3. **What is the focus of the IWRM paradigm?**
   a. Maximizing private water consumption
   b. Minimizing environmental impacts
   c. Enhancing ecosystem services
   d. Balancing social, economic, and environmental needs
4. **Which type of environmental goods is water considered as?**
   a. Private good
   b. Common-pool resource
   c. Club good
   d. Public good
5. **What is the economic management of a river concerned with, in addition to water abstraction?**
   a. Ensuring profitability for private companies
   b. Securing water provision for local ecosystem services
   c. Maximizing water extraction for industrial use
   d. Minimizing government expenditure on water infrastructure
6. **In groundwater management, what does the dynamic stock balance equation represent?**
   a. The equilibrium between water inflow and outflow
   b. The relationship between groundwater extraction and profit
   c. The interaction between groundwater levels and environmental factors
   d. The long-term sustainability of groundwater resources

Answers to Multiple Questions

1. (a); 2. (c); 3. (d); 4. (d); 5. (b); 6. (a)

REFERENCES

1. Al Radif, A. (1999). Integrated water resources management (IWRM): An approach to face the challenges of the next century and to avert future crises. *Desalination, 124*(1–3), 145–153.
2. Allan, J. A. (2006). IWRM: The new sanctioned discourse. *Integrated Water Resources Management: Global Theory, Emerging Practice and Local Needs, 2*(1), 38–63.
3. Allouche, J. (2016). The birth and spread of IWRM–A case study of global policy diffusion and translation. *Water Alternatives, 9*(3), 412.
4. Anderson, A., Karar, E., & Farolfi, S. (2008). Synthesis: IWRM lessons for implementation. *Water SA, 34*(6), 665–669.
5. Arnold, J. G., Allen, P. M., & Bernhardt, G. (1993). A comprehensive surface-groundwater flow model. *Journal of Hydrology, 142*(1–4), 47–69.
6. Asadabadi, E., Asadi, A., & Kalantari, K. (2020). Dynamic modeling of the groundwater system in Hamedan–Bahar Aquifer. *Iranian Journal of Soil and Water Research, 50*(9), 2323–2340.
7. Asher, M. J., Croke, B. F., Jakeman, A. J., & Peeters, L. J. (2015). A review of surrogate models and their application to groundwater modeling. *Water Resources Research, 51*(8), 5957–5973.
8. Balineau, G., & De Melo, J. (2013). Removing barriers to trade on environmental goods: An appraisal. *World Trade Review, 12*(4), 693–718.
9. Ballweber, J. A. (2006). A comparison of IWRM frameworks: The United States and South Africa. *Journal of Contemporary Water Research & Education, 135*(1), 74–79.
10. Barbier, E. B. (2004). Water and economic growth. *Economic Record, 80*(248), 1–16.
11. Brunner, P., Hendricks Franssen, H. J., Kgotlhang, L., Bauer-Gottwein, P., & Kinzelbach, W. (2007). How can remote sensing contribute to groundwater modeling? *Hydrogeology Journal, 15*, 5–18.
12. Cardwell, H. E., Cole, R. A., Cartwright, L. A., & Martin, L. A. (2006). Integrated water resources management: Definitions and conceptual musings. *Journal of Contemporary Water Research & Education, 135*(1), 8–18.
13. Chen, Y., Feng, M., Shi, G., Jiang, M., & Jiang, M. (2023). Bibliometrics-

based research hotspots and development trends in eco-hydrology of dammed rivers. *Chinese Geographical Science, 33*(6), 1153–1164.

14. Cohen, A., & Sullivan, C. A. (2010). Water and poverty in rural China: Developing an instrument to assess the multiple dimensions of water and poverty. *Ecological Economics, 69*(5), 999–1009.
15. Condon, L. E., Kollet, S., Bierkens, M. F., Fogg, G. E., Maxwell, R. M., Hill, M. C., & Abesser, C. (2021). Global groundwater modeling and monitoring: Opportunities and challenges. *Water Resources Research, 57*(12), 29–500.
16. De Alwis, J. M. D. D. J. (2014). Environmental consequence of trade openness for environmental goods. *Sri Lankan Journal of Agricultural Economics, 16*, 79–98.
17. De Graaf, I. D., Sutanudjaja, E. H., Van Beek, L. P. H., & Bierkens, M. F. P. (2015). A high-resolution global-scale groundwater model. *Hydrology and Earth System Sciences, 19*(2), 823–837.
18. De Groot, R. S., Wilson, M. A., & Boumans, R. M. (2002). A typology for the classification, description, and valuation of ecosystem functions, goods, and services. *Ecological Economics, 41*(3), 393–408.
19. Del Moral, L., Pita, M. F., Pedregal, B., Hernández-Mora, N., & Limones, N. (2014). Current paradigms in the management of water: Resulting information needs. In *Progress in Water Geography – Pan-European Discourses, Methods and Practices of Spatial Water Research, Vol 1*, 21–31.
20. Dhungel, R., & Fiedler, F. (2016). Water balance to recharge calculation: Implications for watershed management using systems dynamics approach. *Hydrology, 3*(1), 1–13.
21. Dlamini, C. S. (2012). Types of values and valuation methods for environmental resources: Highlights of key aspects, concepts and approaches in the economic valuation of forest goods and services. *Journal of Horticulture and Forestry, 4*(12), 181–189.
22. Dolan, F., Lamontagne, J., Link, R., Hejazi, M., Reed, P., & Edmonds, J. (2021). Evaluating the economic impact of water scarcity in a changing world. *Nature Communications, 12*(1), 1–10.
23. Eberbach, P. L. (2003). The eco-hydrology of partly cleared, native ecosystems in southern Australia: A review. *Plant and Soil, 257*, 357–369.
24. Fritz, P. (2011). The Factor Four concept in eco-hydrology–Is it applicable? *Ecohydrology & Hydrobiology, 11*(3–4), 141–144.
25. Gallego-Ayala, J. (2013). Trends in integrated water resources management

research: A literature review. *Water Policy, 15*(4), 628–647.

26. Ganoulis, J. (2021). The water–man eristic dialectics for sustainable hydro-governance. *Water International, 46*(7–8), 1135–1157.
27. Gaur, S., Chahar, B. R., & Graillot, D. (2011). Combined use of groundwater modeling and potential zone analysis for management of groundwater. *International Journal of Applied Earth Observation and Geoinformation, 13*(1), 127–139.
28. Grigg, N. S. (2019). IWRM and the nexus approach: Versatile concepts for water resources education. *Journal of Contemporary Water Research & Education, 166*(1), 24–34.
29. Grigg, N. S., & Grigg, N. S. (2016). Framework and scenarios of IWRM. *Integrated Water Resource Management: An Interdisciplinary Approach, 4*(2), 19–31.
30. Hayashi, M., & Rosenberry, D. O. (2002). Effects of groundwater exchange on the hydrology and ecology of surface water. *Groundwater, 40*(3), 309–316.
31. Hellegers, P., Zilberman, D., & Van Ierland, E. (2001). Dynamics of agricultural groundwater extraction. *Ecological Economics, 37*(2), 303–311.
32. Hervani, A. A., Sarkis, J., & Helms, M. M. (2017). Environmental goods valuations for social sustainability: A conceptual framework. *Technological Forecasting and Social Change, 125*, 137–153.
33. Hileman, J., Hicks, P., & Jones, R. (2016). An alternative framework for analyzing and managing conflicts in integrated water resources management (IWRM): Linking theory and practice. *International Journal of Water Resources Development, 32*(5), 675–691.
34. Huang, Y. H., Lai, Y. J., & Wu, J. H. (2022). A system dynamics approach to modeling groundwater dynamics: Case study of the Choshui River basin. *Sustainability, 14*(3), 1371.
35. Isengard, H. D. (2008). Water determination–Scientific and economic dimensions. *Food Chemistry, 106*(4), 1393–1398.
36. Jr Wescoat, J. L. (2013). Reconstructing the duty of water: A study of emergent norms in socio-hydrology. *Hydrology and Earth System Sciences, 17*(12), 4759–4768.
37. Kalantari, K., Maknoon, R., & Karimi, D. (2018). Developing a sustainable legal framework for the establishment of integrated water resources management in Iran. *International Journal of Environmental Research, 12*, 223–231.

38. Kalf, F. R., & Woolley, D. R. (2005). Applicability and methodology of determining sustainable yield in groundwater systems. *Hydrogeology Journal, 13*, 295–312.
39. Kerkhof, A., Drissen, E., Uiterkamp, A. S., & Moll, H. (2010). Valuation of environmental public goods and services at different spatial scales: A review. *Journal of Integrative Environmental Sciences, 7*(2), 125–133.
40. Khan, S., Yufeng, L., & Ahmad, A. (2009). Analyzing complex behavior of hydrological systems through a system dynamics approach. *Environmental Modelling & Software, 24*(12), 1363–1372.
41. Lautze, J., De Silva, S., Giordano, M., & Sanford, L. (2011). Putting the cart before the horse: Water governance and IWRM. *Natural Resources Forum, 35*, 1–8.
42. Li, R., Chen, Q., Tonina, D., & Cai, D. (2015). Effects of upstream reservoir regulation on the hydrological regime and fish habitats of the Lijiang River, China. *Ecological Engineering, 76*, 75–83.
43. Liu, H., Zhang, J., & Lei, H. (2022). Do imported environmental goods reduce pollution intensity? The end use matters. *Energy Economics, 112*, 106–130.
44. Livesey, D. (2010). Measuring the environmental goods and services sector. *Economic & Labor Market Review, 4*, 45–58.
45. Lundqvist, J., Appasamy, P., & Nelliyat, P. (2003). Dimensions and approaches for Third World city water security. *Philosophical Transactions of the Royal Society of London. Series B: Biological Sciences, 358*(1440), 1985–1996.
46. Maxwell, R. M., & Miller, N. L. (2005). Development of a coupled land surface and groundwater model. *Journal of Hydrometeorology, 6*(3), 233–247.
47. Mazvimavi, D., Hoko, Z., Jonker, L., Nhapi, I., & Senzanje, A. (2008). Integrated water resources management (IWRM)–From concept to practice. *Physics and Chemistry of the Earth, Parts A/B/C, 33*(8–13), 609–613.
48. McDonnell, R. A. (2008). Challenges for integrated water resources management: How do we provide the knowledge to support truly integrated thinking? *International Journal of Water Resources Development, 24*(1), 131–143.
49. Mehta, L., & Movik, S. (2014). Flows and practices: Integrated water resources management (IWRM) in African contexts. *IDS Working Papers, 2014*(438), 1–34.

50. Meran, G., Siehlow, M., von Hirschhausen, C., & Meran, G., Siehlow, M., & von Hirschhausen, C. (2021). Integrated water resource management: Principles and applications. *The Economics of Water: Rules and Institutions, 3*(2), 23–121.
51. Merlo, M., Milocco, E., Panting, R., & Virgilietti, P. (2000). Transformation of environmental recreational goods and services provided by forestry into recreational environmental products. *Forest Policy and Economics, 1*(2), 127–138.
52. Merrill, N. H., & Guilfoos, T. (2018). Optimal groundwater extraction under uncertainty and a spatial stock externality. *American Journal of Agricultural Economics, 100*(1), 220–238.
53. Mitchell, B. (2005). Integrated water resource management, institutional arrangements, and land-use planning. *Environment and Planning A, 37*(8), 1335–1352.
54. Moore, C., & Doherty, J. (2006). The cost of uniqueness in groundwater model calibration. *Advances in Water Resources, 29*(4), 605–623.
55. Nuttle, W. K. (2002). Eco-hydrology's past and future in focus. *Eos, Transactions American Geophysical Union, 83*(19), 205–212.
56. Pahl-Wostl, C., Jeffrey, P., Isendahl, N., & Brugnach, M. (2011). Maturing the new water management paradigm: Progressing from aspiration to practice. *Water Resources Management, 25*, 837–856.
57. Rogers, P., De Silva, R., & Bhatia, R. (2002). Water is an economic good: How to use prices to promote equity, efficiency, and sustainability. *Water Policy, 4*(1), 1–17.
58. Rolls, R. J., Leigh, C., & Sheldon, F. (2012). Mechanistic effects of low-flow hydrology on riverine ecosystems: Ecological principles and consequences of alteration. *Freshwater Science, 31*(4), 1163–1186.
59. Rosenberg, D. M., McCully, P., & Pringle, C. M. (2000). Global-scale environmental effects of hydrological alterations: Introduction. *BioScience, 50*(9), 746–751.
60. Roumasset, J., & Wada, C. A. (2013). A dynamic approach to PES pricing and finance for interlinked ecosystem services: Watershed conservation and groundwater management. *Ecological Economics, 87*, 24–33.
61. Saravanan, V. S., McDonald, G. T., & Mollinga, P. P. (2009, February). A critical review of integrated water resources management: Moving beyond polarized discourse. *Natural Resources Forum, 33*(1), 76–86.
62. Savenije, H. H. (2002). Why water is not an ordinary economic good, or

why the girl is special. *Physics and Chemistry of the Earth, Parts A/B/C, 27*(11–22), 741–744.

63. Savenije, H. H., Hoekstra, A. Y., & van der Zaag, P. (2014). Evolving water science in the Anthropocene. *Hydrology and Earth System Sciences, 18*(1), 319–332.
64. Schiemer, F. (2016). Building an eco-hydrological framework for the management of large river systems. *Ecohydrology & Hydrobiology, 16*(1), 19–25.
65. Scibek, J., Allen, D. M., Cannon, A. J., & Whitfield, P. H. (2007). Groundwater–Surface water interaction under scenarios of climate change using a high-resolution transient groundwater model. *Journal of Hydrology, 333*(2–4), 165–181.
66. Shang, S., Gabriel, H. F., & Zhang, Q. (2024). Editorial on Hydrology and Water Resources in Agriculture and Ecology. *Remote Sensing, 16*(2), 238.
67. Siwar, C., & Ahmed, F. (2014). Concepts, dimensions, and elements of water security. *Pakistan Journal of Nutrition, 13*(5), 1–281.
68. Stenger, A., Harou, P., & Navrud, S. (2009). Valuing environmental goods and services derived from the forests. *Journal of Forest Economics, 15*(1–2), 1–14.
69. Tague, C. L., Moritz, M., & Hanan, E. (2019). The changing water cycle: The eco-hydrologic impacts of forest density reduction in Mediterranean (seasonally dry) regions. *Wiley Interdisciplinary Reviews: Water, 6*(4), e1350.
70. Thomas, J. S., & Durham, B. (2003). Integrated water resource management: Looking at the whole picture. *Desalination, 156*(1–3), 21–28.
71. Tremblay, H. (2011). Clash of paradigms in the water sector-tensions and synergies between integrated water resources management and the human rights-based approach to development. *Natural Resources, 51*, 307.
72. Wagner, I., & Breil, P. (2013). The role of ecohydrology in creating more resilient cities. *Ecohydrology & Hydrobiology, 13*(2), 113–134.
73. Whitehead, J. C., & Blomquist, G. C. (1991). Measuring contingent values for wetlands: Effects of information about related environmental goods. *Water Resources Research, 27*(10), 2523–2531.
74. Zalewski, M. (2015). Ecohydrology and hydrologic engineering: Regulation of hydrology-biota interactions for sustainability. *Journal of Hydrologic Engineering, 20*(1), A401–A4012.
75. Zhou, Y., & Li, W. (2011). A review of regional groundwater flow modeling. *Geoscience Frontiers, 2*(2), 205–214.

CHAPTER 8

CLIMATE CHANGE IMPACT AND GIS APPLICATIONS IN WATER RESOURCES

LEARNING OBJECTIVES

At the end of this chapter, readers will be able to:

1. Understand climate change and its attribution to human and natural factors.
2. Know radiative forcing and natural processes influencing climate change.
3. Understand the impacts of climate change on water resources including runoff and floods.
4. Measure the effects of climate change on drought occurrence and severity.
5. Know the implications of climate change on snowmelt, glacier retreat, and associated runoff.
6. Examine the influence of climate change on water quality and groundwater resources.
7. Understand GIS applications in water supply system design and operation.
8. Explore GIS techniques for groundwater assessment and modeling.
9. Learn about GIS utilization in floodplain mapping and hydraulic modeling.

INTRODUCTORY EXAMPLE

Coca-Cola, a multinational corporation renowned for its beverages, confronts the repercussions of climate change on its water management strategies. Operating across diverse regions worldwide, the company faces mounting challenges in areas susceptible to water scarcity and unpredictable weather conditions. Notably, regions like California in the United States and various parts of India grapple with prolonged droughts, compelling Coca-Cola to adapt its water sourcing practices to ensure sustainability.

To address these challenges, Coca-Cola is actively implementing water conservation measures and forging partnerships with local communities and authorities. In drought-affected regions, the company invests in water-efficient technologies and wastewater recycling facilities to minimize environmental impact while maintaining operational continuity. Moreover, Coca-Cola recognizes the importance of integrating advanced technologies, such as Geographic Information Systems (GIS), into its water management approach.

By harnessing GIS technology, Coca-Cola gains valuable insights into water availability, quality, and consumption patterns across its global operations. This data-driven approach enables the company to identify vulnerable areas, assess risks associated with climate change, and develop targeted strategies to enhance resilience and sustainability. Through proactive measures and innovative solutions, Coca-Cola exemplifies how corporations can navigate the challenges posed by climate change while fostering responsible water stewardship for future generations.

UNIT INTRODUCTION

Water is essential for the existence and endurance of all living organisms. The availability of freshwater resources is finite, thus necessitating the most significance of protecting and overseeing their preservation and administration. Understanding the relationships between freshwater, climate, and socioeconomic and biophysical systems at many different geographic scales including watershed, regional, and global scales is essential to the sustainable management of freshwater resources. A major threat to the availability of high-quality fresh water, as well as to the amount (too much or too little), has been recently documented as behaviors that contribute to climate change (Braca et al., 2019). This chapter delves into the complex relationship between climate change and water supplies, as well as the novel solutions provided by

Geographic Information Systems (GIS). The chapter begins by exploring the basics of climate change, including its defining features and the various factors that contribute to its occurrence. Next, the factors behind climate change, including the impact of greenhouse gas emissions and various environmental dynamics such as radiative forcing and natural processes will be discussed (Singh & Gosain, 2011).

The climate change effects on water resources are then examined, including modified runoff patterns, increased flood hazards, worsened drought conditions, snowmelt, and glacier melt, weakened water quality, and the complex disturbances caused to groundwater reserves and ecosystems.

The critical role that GIS plays in flood mitigation and water resource management. GIS shows up as a potent tool for boosting resilience against climate-induced issues, from enabling robust modeling and analysis in groundwater hydrology to optimizing water supply systems through demand forecasts and network design will also be explored (Wang & Xie, 2018).

8.1. CLIMATE CHANGE

Learning Objectives:

- Understand climate change and its persistence over extended periods.
- Identify anthropogenic and natural factors contributing to climate change.

8.1.1. Definition of Climate Change

Understanding the climatic patterns of a region is crucial for comprehending the functioning of natural ecosystems. These patterns can be measured by analyzing hydro meteorological variables like temperature, precipitation, humidity, and wind. The Intergovernmental Panel on Climate Change (IPCC) offers a thorough explanation of climate change, describing it as persistent changes in the characteristics of the climate. The observed changes can be linked to a range of variables, encompassing natural internal processes, external forcing, and enduring alterations caused by human activities in the composition of the atmosphere or land use (Pielke, 2004).

The United Nations Framework Convention on Climate Change (UNFCCC) defines climate change that emphasizes the impact of human activity on the global atmosphere, besides the natural climate variability observed over similar periods. Climate change, as per the definition provided by the AMS Glossary, refers to a persistent alteration over a period of several decades or longer in the statistical patterns of climate variables, such as pressure, temperature, or winds. Climate change is caused by a variety of reasons, including both internal climate system processes and external pressures like variations in solar output or changes in the components that make up Earth's orbit. In addition, human activities also have a significant impact on driving climate change (Berrang et al., 2011).

8.1.2. Causes of Climate Change

Climate change is a result of various factors that interact with each other, including the Earth, atmosphere, ocean, and land systems. These changes can arise from either natural processes or human activity, such as alterations in greenhouse gas concentrations, levels of aerosols, land cover, land use, and solar radiation. These factors influence the manner in which radiation is absorbed, scattered, and released in both the atmosphere and on the Earth's surface (Cahill et al., 2013). A few significant elements that contribute to climate change will be examined in subsections (Figure 8.1).

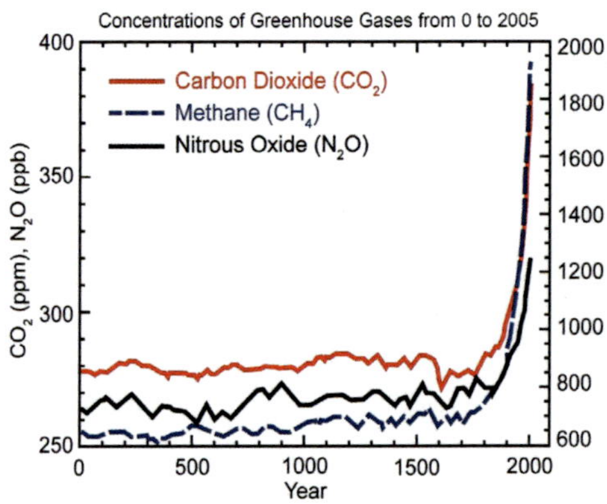

Figure 8.1. *Descriptions of significant long-lived greenhouse gas concentrations in the atmosphere during the previous 2,000 years. Concentration units are parts per billion (ppb) and parts per million (ppm).*

Source: H. Lins; Creative Commons License.

8.1.2.1. Greenhouse Gases

Alterations in the Earth's atmosphere, including fluctuations in greenhouse gases, aerosols, and cloud cover, play a crucial role in the phenomenon of climate change. The rise in global temperatures in recent years can be primarily attributed to the significant increase in greenhouse gas emissions resulting from human activities, rather than natural fluctuations. During the industrial era, human activities have greatly influenced the Earth's climate, resulting in a noticeable rise in global temperatures (Montzka et al., 2011).

The main greenhouse gases, such as methane (CH_4), carbon dioxide (CO_2), the halocarbons, and nitrous oxide (N_2O), are produced by human activities. These gases build up in the atmosphere and their concentration increases over time, as shown in Figure 8.1. There has been a significant rise in CO_2 levels, primarily due to the extensive utilization of fossil fuels in building cooling and heating, transportation, and the production of cement and other commodities.

Various human activities, including natural gas distribution, agriculture, and landfills, contribute to the increase in CH_4 emissions. In addition, the application of fertilizer and the combustion of fossil fuels both contribute to the increase in N_2O levels. It has been discovered that the growing utilization of chlorofluorocarbons, commonly known as principal halocarbons, in refrigeration and other industrial applications is leading to the depletion of stratospheric ozone (Ledley et al., 1999).

8.1.2.2. Radiative Forcing

Radiative forcing, which is measured at the apex of the atmosphere, indicates the rate of energy change per unit area of the Earth. It is used to analyze the energy balance of the Earth-atmosphere system. When there is positive radiative forcing, the Earth-atmosphere system becomes warmer. Conversely, negative radiative forcing causes the energy to decrease, resulting in a cooling of the system (Hansen et al., 1997).

Various factors contribute to forcing, including the increase in greenhouse gases, the warming effect of tropospheric ozone, the cooling effect of stratospheric ozone depletion, the influence of aerosol particles through reflection and absorption, changes in land cover such as pastures, croplands, and forests, and the presence of condensation trails caused by aircraft under specific weather conditions.

Did You Know?

Extreme rainfall events are expected to occur more frequently and intensely in many areas due to climate change.

8.1.2.3. Natural Processes

These days, the influence of human activity on climate greatly outweighs that of natural phenomena like variations in the sun's surface and volcanic eruptions. Three types of orbital fluctuation are discussed in the original Milankovitch theory that may have an impact on climate patterns: the precession of the equinoxes, the tilt of the Earth's axis, and the eccentricity of the Earth's orbit around the Sun (Swanston, 1991).

Large volumes of gases can be released during volcanic eruptions, which may affect how much solar radiation reaches the surface of the Earth. Changes in atmospheric and temperature circulation patterns may result from this. Moreover, air circulation and greenhouse feedback may be impacted by tectonic motions directly or indirectly (Paola et al., 2011).

8.2. CLIMATE CHANGE IMPACT ON WATER RESOURCES

Learning Objectives

- Understand the impact of climate change on runoff patterns and flooding.
- Understand the climate change effects on drought occurrence and severity.
- Know the consequences of climate change on snowmelt and glacier retreat.
- Analyze how climate change influences water quality and groundwater resources.
- Explore the implications of climate change on ecosystems and biodiversity.

Water supplies are significantly impacted by changes in climatic trends. The section that follows discusses the likely consequences of climate change on the water resources industry.

8.2.1. Runoff

Climate change will have a significant impact on the hydrological system, changing river flows and local water supplies based on existing measurements and climatic projections. The variability in weather patterns results in the occurrence of floods in various regions and at varying intervals. In river discharge analysis, numerous studies conducted in the 20th century explored potential trends in various indicators of flow at different spatial scales. Some of these studies successfully identified significant trends in flow measures, while others established statistically significant connections between these trends and changes in temperature or precipitation (Tholin & Keifer, 1960).

Furthermore, the actions of humans have had a significant impact on the flow patterns in various catchments worldwide. This has resulted in a noticeable and consistent shift in the annual runoff. Certain areas, like high latitudes and large portions of the USA, have witnessed an increase in runoff. Conversely, regions such as southern Europe, West Africa, and southernmost South America parts have observed a decrease in runoff (Jakeman & Hornberger, 1993).

8.2.2. Floods

Floods have a devastating impact on the economies of affected regions, making them one of the most frequent natural disasters worldwide, largely influenced by climate change. Multiple factors contribute to flooding, such as intense rainfall, torrential downpours, and the melting of snow. The spatial distribution of these events since 1985 is shown in Figure 8.2.

Severe floods caused by heavy rainfall, whether it be short or long in duration, have been observed in various humid regions across the globe, and certain semiarid areas. Tropical storms also referred to as cyclones, hurricanes, or typhoons, have a more concentrated distribution. These weather phenomena tend to occur in specific hotspots, such as the Caribbean, southeastern USA, western Pacific coasts, and the Bay of Bengal. Flooding in India is often caused by the heavy rainfall during the monsoon season, leading to the breaking of levees. In August 2002, a major event of severe flooding took place in five countries of central and Eastern Europe. This was caused by prolonged and heavy rainfall (Merz et al., 2021).

Identifying the impacts of human activities on flood patterns can be quite challenging, primarily because of the significant natural fluctuations. Additionally, changes in land use can further complicate the matter by affecting the flow of water. As an illustration, there has been a significant rise in the occurrence of major floods, specifically those with discharges surpassing a 100-year return period from basins larger than 2,00,000 km^2. The increase has been established by combining measurements of stream flow with computational predictions of climate change caused by human activities. Notably, the frequency of these major floods has seen a substantial surge throughout the 20th century (Merz & Blöschl, 2003).

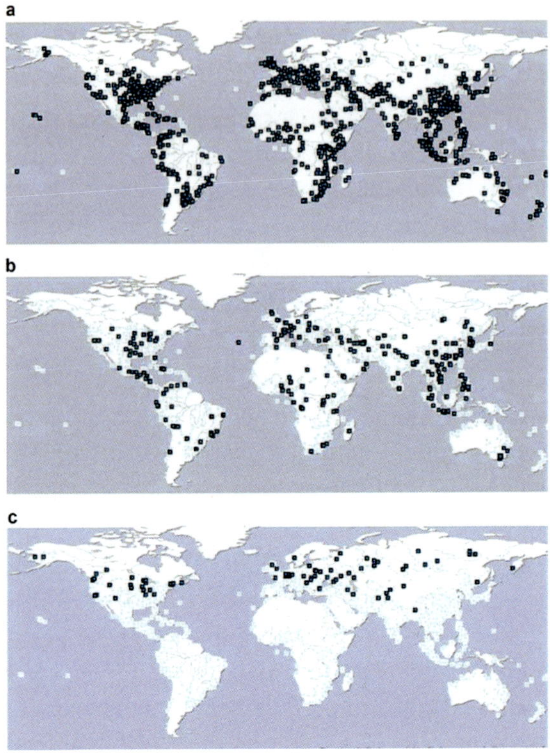

Figure 8.2. *Illustration of the Dartmouth Flood Observatory that has documented the spatial distribution of catastrophic floods from three different causes. (a) Heavy rain; (b) brief torrential rain; and (c) snowmelt.*

Source: E. Stakhiv; Creative Commons License.

The number of catastrophic inland flood disasters worldwide has increased significantly during the last 10 years, surpassing the numbers from 1950 to 1980. Additionally, the economic losses associated with these floods have multiplied by a factor of five. Nevertheless, in a climate that is becoming warmer and more variable, the possibility of flooding is expected to rise (Costa & O'Connor, 1995).

8.2.3. Drought

Throughout the 20th century, natural hazards have caused significant damage, with droughts being particularly devastating. In recent decades, extensive and severe droughts have occurred on every continent, impacting vast regions in Africa, Asia, Europe, South America, Australia, Central America, and North

America (Mishra & Singh, 2010). For the last two centuries, Western Canada has experienced 40 long-duration droughts. Multiyear droughts were observed in the southern regions of Alberta, Saskatchewan, and Manitoba during the 1890s, 1930s, and 1980s. The severity of the drought situation in several European regions has escalated. Over the past 30 years, Europe has experienced several significant drought events. These include the droughts in 1976, 1989, and 1991, and the more recent prolonged drought in 2003, which was accompanied by a summer heat wave (Wilhite & Glantz, 1985).

The frequency and severity of droughts in the USA have led to a significant increase in their impacts. According to the data provided by the NCDC, USA (2002), a significant portion of the United States' land area has been affected by severe or extreme drought at various points throughout the past century.

Did You Know?

Droughts not only affect water availability but also have far-reaching ecological and socio-economic impacts, including crop failures, water shortages for agriculture, and disruptions to ecosystems.

The occurrence of frequent and severe droughts in 1997, 1999, and 2002 in various regions of Northern China resulted in substantial economic and societal damages. The 1997 drought in Northern China caused the Yellow River to experience a 226-day stretch without any streamflow, making it the longest recorded time of drying up. Since the late 1970s, the progression of global warming has led to greater temperatures and increasing drying, resulting in an elevated risk of droughts. Furthermore, Australia experiences a periodic pattern of drought, with the most recent and prolonged one being the millennium drought, which has persisted for nearly 10 years (Diaz, 1983).

Several researchers have emphasized an increase in drought episodes in the 21st century. The majority of the Earth's land is expected to see a significant decrease in moisture during the non-growing season, except for some areas in the northern mid- and high-latitude regions. This prediction is based on the analysis of 15 coupled models from the IPCC AR4 simulations under the SRES A1B scenario (Farooq et al., 2009). This suggests that by the end of the 21st century, there will be a worldwide agricultural drought. The findings from the eight AR4 models indicate a concerning trend in soil moisture. Regardless of the scenarios considered, all models project a decrease in soil moisture. This decrease is accompanied by an alarming increase in the severity and frequency of short-term droughts, lasting between 4 and 6 months.

Furthermore, the models predict that by the end of the 21st century, the frequency of long-term droughts lasting longer than a year will have increased by three times. An analysis was conducted to determine the spatial distribution of drought. An analysis was conducted on the trend of the Palmer drought severity index, using the Penman-Monteith potential evapotranspiration as the drought index (PDSI-PM) (Panu & Sharma, 2002). The analysis was performed for each decade at every point using the A2 ensemble (Figure 8.3).

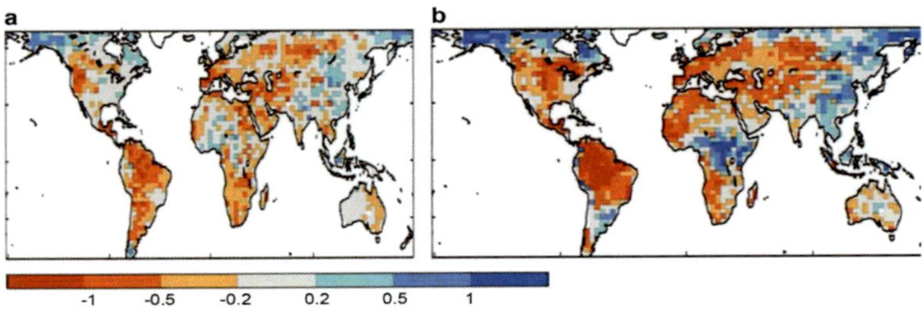

Figure 8.3. *Illustration of trend in PDSI-PM per decade for (a) ensemble mean of the first half of the 21st century; and (b) ensemble mean of the second half of the 21st century projected by SRES A2.*

Source: M. Ayers; Creative Commons License.

According to the observation, there are predictions of drying in certain regions such as the United States, Northern Africa, Amazonia, Southern Europe, and Western Eurasia. On the other hand, there are predictions of wetting in Eastern Asia, Central Africa, and high northern latitudes.

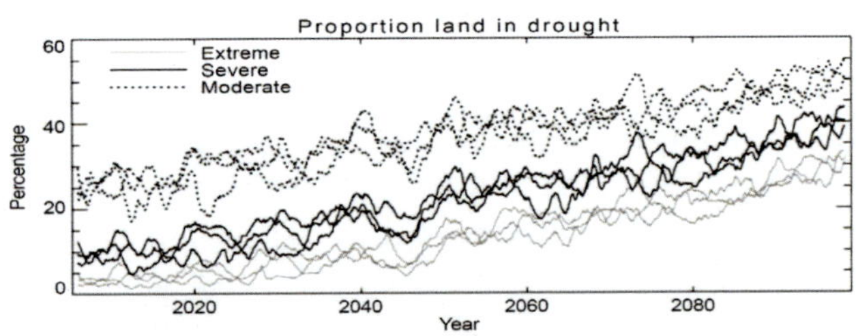

Figure 8.4. *Illustration of land surface proportion in drought for 21st century based on results from A2 emissions scenario.*

Source: Bultot, L.; Creative Commons License.

With a predicted drop in the global average PDSI of 0.30 per decade throughout the first half of the 21st century, there is an overall pattern of drying. In the second half of the century, there was a noticeable increase in the drying trend, as indicated by the PDSI-PM decreasing by 0.56 per decade.

There is a forecasted rise in the percentage of land area experiencing drought conditions over the course of the 21st century. This increase is expected to persist for the whole century, as shown in Figure 8.4. The reader may refer to this source to enhance their comprehension of drought topics and to explore the implications of drought under climate change scenarios (Ault, 2020).

8.2.4. Snowmelt and Glacier Melt

Several significant river systems worldwide receive water from snowpack and melting glaciers. It is highly probable that global warming will significantly affect the process of snowmelt and the resulting discharge. Glaciers are highly responsive to all hydrological factors, such as humidity, precipitation, and wind speed, but their sensitivity is mostly influenced by temperature. As a result, glaciers serve as a reliable indicator of global warming. Glacier retreat is evident on all continents, and it is noticeably influenced by global warming (Singh & Kumar, 1997).

Rivers located in regions with high latitudes and high altitudes may undergo an augmentation in water flow, even if there is a reduction in rainfall, due to the melting of glaciers. Variations in volume and timing of water flow can lead to notable alterations in the physical features of rivers, such as the widening and deepening of the channel, greater winding, enhanced erosion of riverbanks, and accelerated movement of the channel.

The surface area of the European Alps has decreased by approximately 30 to 40%, and its volume has decreased by about half since the Little Ice Age. This decline has been more rapid in the late 20th century. During the period from 1992 to 1998, there was a significant decrease in the volume and thickness of the Bolivian Andes. If this trend continues, the Bolivian Andes could completely disappear over the next 15 years (Schaner et al., 2012).

Three glaciers in North America, including the South Cascade Glacier in the Pacific Northwest and the Wolverine and Gulkana glaciers in Alaska, have exhibited fluctuating changes in their size. In the Himalayas, glaciers have been documented to either retreat or advance at different rates, depending on their location and the historical period. However, the system has experienced a significant general decline in both the extent and thickness of glaciers during the past century (Dietz et al., 2020). The Gangotri Glacier, which serves as the source of the Ganges River, has experienced a reduction in size of 1.5 km over a span of 69 years. However, its current retreat is occurring at a slower pace. The Alaskan glacier had a significant decline in size between the mid-1950s and the mid-1990s, which caused the sea level to rise by about 0.14 to 0.04 mm/year.

This shrinkage amounted to a loss of 52.15 km^3/year. During the period from the mid-1990s to 2000–2001, there has been a significant acceleration in the thinning of Alaskan glaciers. This has resulted in a substantial volume loss of 96 35 km^3 per year and a corresponding rise in sea levels of 0.27 0.10 mm per year. During the time span from 1961 to 1997, sub-polar glaciers experienced an average annual shrinkage of 147 mm. This significant reduction in size amounted to a total volume loss of 3.7 103 km^3. The estimated sea level rise for the period 1961–2003 was 0.51 mm/year, with a faster rate of 0.93 mm/year observed from 1994–2003, indicating accelerated melting (Mukhopadhyay & Khan, 2015).

Quantifying the societal impacts of retreating glaciers can be quite challenging. Glacial melt plays a crucial role in supporting various aspects of human life, such as agriculture, domestic water supply, and hydroelectric power generation, through the nourishment of major rivers. Without the necessary infrastructure, the increased water from melting glaciers could result in catastrophic floods or droughts due to changes in the timing of the runoff. The projected expenses for acquiring extra power generation capacity to mitigate the consequences of receding glaciers in Peru amount to almost US $1 billion per gigawatt (Omani et al., 2016).

8.2.5. Water Quality

Over the past century, there has been a significant increase in temperature, and climate change models are predicting further temperature rises. Higher temperatures have various impacts on water quality, such as lower levels of dissolved oxygen, higher amounts of contaminants entering water bodies, decreased flow in streams and rivers, more frequent algal blooms, and a greater risk of saltwater intrusion in coastal areas (Aston, 1973).

One of the main factors contributing to the decline in water quality is acid rain, which is primarily caused by nitrogen and sulfuric compounds resulting from human activities. Additionally, changes in stream flow patterns can impact the movement of chemical loads within rivers. As a result of decreased river flows, the interaction between water and waste will be disrupted, potentially leading to an intensified mineralization of organic nitrogen in the soil (Gorde & Jadhav, 2013).

Higher water temperatures result in reduced oxygen levels, making global warming a factor in decreasing dissolved oxygen content. Greater precipitation intensity leads to an increase in runoff and erosion, which causes a higher level of pollutant transport. Many elements, including the makeup of the soil, geological formations, and the topography of catchment areas, can have an impact on the quality of the water. Human activities, precipitation, surrounding vegetation, and runoff from adjacent land also play a role. Additionally, biological, physical, and chemical processes in water contribute to its overall quality (Bhateria & Jain, 2016).

Research findings indicate an increase in surface water temperature in North America, Europe, and Asia since the 1960s, ranging from 0.2 to 2°C. This rise can be attributed primarily to atmospheric warming caused by an increase in solar radiation. After the severe drought of 2003, the Rhine and Meuse Rivers had a notable rise in water temperature of around 2°C, which therefore resulted in a drop in the quantities of dissolved oxygen (DO). There has been an extension of the stratified period by 2–3 weeks in various lakes in Europe and Northern America. This is a result of a temperature increase of 0.2–1.5°C, which has impacted thermal stratification and Lake Hydrodynamics.

Various factors that are impacted by climate change, such as changes in rainfall intensity, levels of atmospheric CO_2, air temperature, and acid deposition, have resulted in notable increases in dissolved organic carbon (DOC) in Central Europe, Northern Europe, and North America (Beck, 1987).

Weather pattern fluctuations have a notable influence on nutrient loading. The warming climate has a significant impact on water bodies, resulting in higher levels of nutrients in both surface and groundwater. Increased air temperatures and variations in rainfall patterns, including seasonality and intensity, are the main factors influencing the shifting fate and behaviors of pesticides (Hertzberg et al., 2017).

Numerous research papers have delved into the effects of climate change on water quality in the coming decades. By 2070, European lakes could face a significant challenge due to climate change, particularly shallow lakes. The risk of adverse effects will be influenced by factors such as water temperature, lake characteristics, and the season. Furthermore, it is projected that in 2050, lakes with short residence times may experience a significant increase of 92% in their residence time during the summer months.

Over an extended period, lakes with greater heat storage capacity are more susceptible to climate warming, resulting in higher winter temperatures. Changes in water temperature can have a significant impact on the chemical processes occurring in lakes, leading to alterations in pH levels and an increase in alkalinity within the lake. Rivers may have higher nitrate concentrations in the summer because of an increase in the frequency of incidents (Omer, 2019).

There is a possibility that nitrates could be gradually released from soils and transported into streams during the onset of the wet seasons. Past studies have discovered an elevation in nitrate levels within the aquifer layers of the Seine basin, as well as a notable surge in nitrate flow within a river basin in Norway. These changes are attributed to higher levels of precipitation, which in turn lead to increased soil leaching (Edition, 2011).

8.2.6. Groundwater

The global warming impact on groundwater resources is expected to be significant, as it will disrupt precipitation and temperature patterns. This issue may be exacerbated by the excessive exploitation of these resources. India's states of Rajasthan, Punjab, and Haryana are seeing a 33 cm annual decline in groundwater levels, according to data from NASA's Gravity Recovery and Climate Experiment (GRACE) twin satellites (Sophocleous, 2002).

Studies conducted in the United States have revealed a concerning trend in the San Joaquin Valley in California. Over the past 5 years, there has been a significant decline in groundwater levels, with a drop of 60–150 cm. The recharge of groundwater can be influenced by various factors including alterations in land use and land cover, urban development, deforestation, modifications in crop rotation and patterns, and modifications in soil properties. These changes can occur gradually over an extended period and have the potential to impact the capacity of infiltration (Konikow & Kendy, 2005).

Based on the projected climate change and water demands for the year 2050, the Edwards Balcones Fault Zone Aquifer in Texas, USA is expected to experience changes in its recharge dynamics. These changes are likely to lead to an increase in the spring flow in the area. However, the growing demand for water will be the primary concern for the aquifer (Giordano, 2009).

An investigation into the impact of increased CO_2 levels on the hydrological cycle in Bie´vre-Valloire, Grenoble, revealed that while global warming is not expected to significantly alter rainfall patterns, it could lead to a substantial rise in evaporation and a decline in groundwater recharge. In addition, alterations in land use and precipitation patterns, such as sudden heavy rainfall instead of light drizzles, have the potential to modify runoff patterns and decrease recharge even more.

In several regions, including southwest Africa, northeastern Brazil, and the southern rim of the Mediterranean Sea, the computed groundwater recharge experiences a significant decrease of over 70% based on the four climate scenarios. By the 2050s, several regions are projected to experience a significant increase in groundwater recharge. These areas include the Sahel, Northern China, the Middle East, the western USA, and Siberia (Wada et al., 2010).

Under the Pressures of Humanity and Climate Changes (GRAPHIC) project, the UNESCO-International Hydrological Program (IHP) has launched the Groundwater Resources Assessment to examine the potential future difficulties associated with groundwater resources. This project aims to investigate the relationship between the global water cycle and groundwater, its role in sustaining ecosystems and human life, and the risks posed by population growth and climate change. The research will be conducted across all continents, utilizing a range of cutting-edge technologies to evaluate and track alterations in groundwater levels, such as the implementation of GRACE (Tóth, 1999).

Practice Problem

WARSA Water Management Company, operating in a region prone to groundwater depletion, faces the challenge of assessing the potential impacts of climate change on groundwater resources. With historical data indicating excessive pumping for agricultural and urban use leading to declining groundwater levels, the company anticipates that projected decreases in precipitation due to climate change will exacerbate the problem. As a result, there is an urgent need to evaluate the future availability of groundwater and devise effective strategies to mitigate these impacts while ensuring sustainable water management practices.

Solutions to Practice Problem

Leveraging their expertise in Geographic Information Systems (GIS) and water resource management, WARSA Water Management Company embarks on a comprehensive assessment of the situation. They utilize GIS technology to map out current groundwater levels, recharge rates, and extraction patterns in the region. Incorporating climate change projections, they model future groundwater scenarios to visualize potential impacts. With these insights, the company develops tailored mitigation strategies, including promoting water conservation measures, implementing efficient irrigation techniques, and exploring options for artificial recharge of aquifers. By fostering collaboration with stakeholders and continuously monitoring groundwater dynamics using GIS tools, WARSA ensures adaptive management practices to safeguard groundwater resources in the face of climate change challenges.

8.2.7. Ecosystems

The rapid rise in the planet's energy balance due to global climate change has caused major changes in physical processes and biological systems at different scales. These changes have had a profound impact on ecosystems that play a crucial role in supporting human society. The potential consequences of this phenomenon are significant, as they can lead to abrupt and permanent alterations in the ecosystem's function and structure. These changes have the potential to greatly impact human society (Jacobides et al., 2018). The global ecosystem is influenced by various factors, including significant fluctuations and extremes in CO_2 levels and climate over time. Additionally, human activities such as nitrogen deposition, land use, pollution, and the introduction of invasive species have an impact. Natural disturbances like wildfires and even small changes in management practices within specific land-use types, such as intensified agricultural practices, also play a role.

The impact of habitat loss, land-use changes, and fragmentation on ecosystem change has been widely recognized, especially in terms of their effects on biodiversity. Ecosystems can undergo a significant shift when there are slight alterations in physical conditions. These shifts may not be easily reversible and can have asymmetrical effects. Initially, ecosystems tend to resist change, but they can respond in a nonlinear way when faced with external forcing (Hannon, 1973). Deserts are expected to undergo an increase in occasional climate events and year-to-year fluctuations in the future, primarily because they are one of the largest land ecosystems. Nevertheless, there is substantial variance within GCM projections and across various locations about this issue. The vulnerability to desertification is expected to worsen as severe droughts become more frequent worldwide. For instance, temperate deserts in the Americas are predicted to significantly expand under climate scenarios with doubled levels. In addition, the impact of climate change on desert biodiversity is a cause for concern. For example, the Succulent Karoo biome in South Africa, which currently provides a suitable habitat for around 2,800 plant species, could potentially see an 80% reduction in its biodiversity if global temperatures rise by 1.5–2.7°C above preindustrial levels. The impact of changes in rainfall patterns and variability on vegetation in tropical grassland and savanna systems is highly probable (Pimm, 1984).

Given that changes in rainfall patterns and amounts are anticipated, temperate grasslands may react differently to future climate change. In the regions of the Mediterranean Basin, there may be a rise in the frequency of fires. According to climate scenarios doubled levels, the occurrence of wildfires could increase by 40–50% in California and the fire risk in Cape Fynbos could double. Climate change will have a direct impact on inland aquatic ecosystems, as temperatures rise and concentrations increase. Additionally, changes in global or regional precipitation patterns will indirectly affect these ecosystems by altering the hydrology. Higher temperatures will have a negative impact on microorganisms, benthic invertebrates, and various species of fish. However, the shift in poleward of invertebrates, waterfowl, and tropical invasive biota may lead to the extinction of some species (Patten, 1981).

There are expected to be significant shifts in the types of species present, the timing of their life cycles, and the overall dynamics of planktonic communities and their interactions within the food web. The concentration of dissolved organic carbon will be significantly affected by the combination of enhanced UV-B radiation and increased summer precipitation, leading to significant changes in major biogeochemical cycles. The variability of precipitation in monsoonal regions has a significant impact on the plants and animals that inhabit wetlands. Even small increases in this variability can have detrimental effects on wetland biodiversity. Additionally, prolonged dry periods can lead to the transformation of wetlands into terrestrial areas, as seen in Keoladeo National Park, India (Da Silva & Wheeler, 2017). Climate change and shifting land-use patterns are putting more strain on the riparian ecosystems of many rivers worldwide. These changes in freshwater flows will have a significant impact on coastal wetlands, affecting factors such as sediment inputs, salinity, and nutrient loadings (Brundrett, 1991).

8.3. APPLICATION OF GIS IN WATER RESOURCES AND FLOOD

Learning Objectives:

- Understand how GIS supports water supply system design and operation.
- Understand the role of GIS in groundwater assessments and modeling.
- Know about GIS applications in floodplain mapping and hydraulic modeling.
- Understand FEMA standards and geodatabase integration in floodplain management.

8.3.1. GIS for Water Supply Systems

8.3.1.1. Overview of Water Supply Systems

Water supply systems play a crucial role in supporting the well-being and economic development of communities. This chapter provides an overview of water supply distribution systems and their various components. It also explores the design and operation aspects of these systems, along with the use of GIS concepts and tools to support these activities. GIS tools are commonly used to estimate demand, design networks, and manage system operations. These functions are made possible using GIS spatial data management and analysis tools (Sapkota et al., 2014).

When GIS is not utilized, the necessary system parameters are frequently generalized. When it comes to pipe connections, the spatial details are often simplified to a single value that represents the average tendency of a group of connections. However, this approach can lead to significant errors. A GIS offers various functions to develop and prepare precise spatial information for inputting into network design simulation models and operations control (Kiselychnyk et al., 2009).

8.3.1.2. GIS-Based Water Supply Demand Forecasting

Wu, Prins, and Bordeaux gave directions on how to generate water demands for GIS-based pipe network modeling. Water demands are typically determined by combining land use maps with attribute data (e.g., land use category, customer ID, water consumption records, and per unit planning variables) from a relational database, or geodatabase (Jayarathna et al., 2017).

Historic customer billing records can provide valuable insights into water demands. Geocoding, a common GIS process for determining customer locations based on their addresses, can be utilized to allocate customer demands to a model using different methods: nearest pipe, nearest node, or meter aggregation. Figure 8.5 provides a visual representation of how demand is assigned to different nodes based on land use (Masera et al., 2006).

8.3.1.3. Pipe Network Design with GIS

Hydraulic models are used in the design of a water supply distribution system to simulate the network's performance under various design conditions. These scenarios usually include forecasted maximum day and maximum hour demands, as well as fire flow demands. The process of designing a pipe network is iterative, requiring the specification of a layout that includes pipes of specific types and sizes. To ensure reliability, it is important to have a looped system for the network. This allows water to be distributed to all services, even if a particular pipe needs to be shut off for repairs (Ayad et al., 2018).

The layout of the system can be greatly influenced by the topography, as it allows for the utilization of gravity distribution, which is both more reliable and cost-effective compared to pumping. Strategically placing distribution storage tanks in the system serves multiple purposes. It helps accommodate changes in water usage, ensures a water supply for firefighting, and maintains stable pressures in the distribution system. In addition, it may be necessary to utilize pumps to transfer water to elevated storage locations. Achieving an accurate depiction of the hydraulics of the distribution system requires calibrating the hydraulic model (Inanloo et al., 2016).

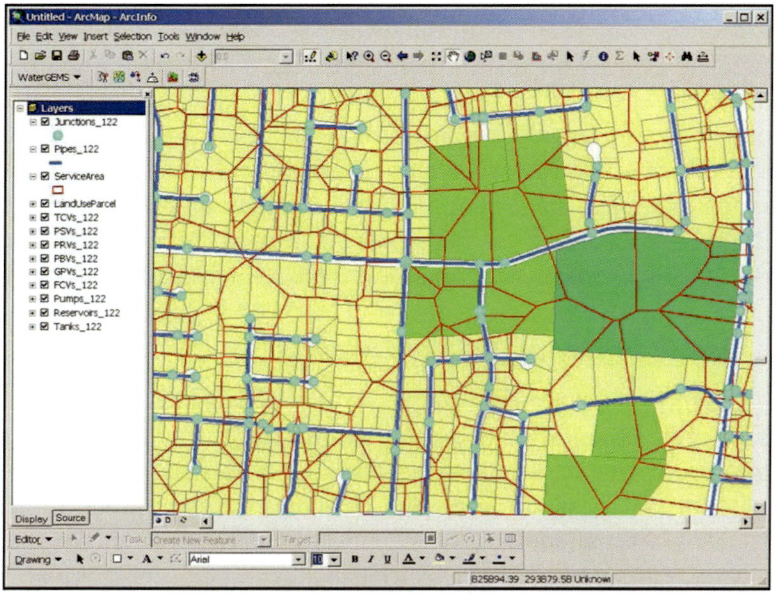

Figure 8.5. *Illustration of the nodal demand assignments by land use example.*

Source: P. Mosley; Creative Commons License.

Pipe network models utilize hydraulic flow and network theory to accurately depict flows and friction losses within a network, employing the principles of conservation of energy and mass. Pipe networks offer complex challenges, especially when the network comprises numerous pipes, and resolving these challenges necessitates advanced computerized mathematical algorithms.

GIS provides advanced features for creating and organizing accurate spatial data to be utilized in the network design process. This includes tasks such as determining network layout, establishing connections, defining pipe properties, calculating pressure changes, analyzing costs, determining optimal routes, allocating resources, and visually presenting results using color graphics. The GIS system handles various tasks related to database administration, including managing attribute and spatial data. It also offers user-friendly interfaces for modifying data and presenting outputs. Additionally, it features a model subsystem that covers both simulation and optimization (Haywood & Boucher, 2000).

8.3.2. GIS in Groundwater Hydrology

8.3.2.1. Overview of GIS in Groundwater

GIS has been widely used for groundwater assessments due to the variety and volume of data involved. Thorough geologic, hydrologic, and hydraulic investigations are essential for accurately evaluating groundwater resources. The spatial range can vary greatly, ranging from a small pumping well to areas that can span from just a few 100 acres to entire basins or even entire countries. The use of simulation and management models is frequent in these types of studies, with GIS serving as a crucial technology for data coordination and providing an interface for the development of groundwater models (Elbeih, 2015).

8.3.2.2. Groundwater and GIS Modeling

Utilizing techniques for groundwater modeling, researchers can replicate field data and examine how the groundwater system reacts to changes in land use, population expansion, water demand trends, and climate variables. The US Geological Survey created the modular finite-difference groundwater flow model MODFLOW, which is a widely used computer model for numerical analysis. MODFLOW is a powerful tool that accurately models the flow of groundwater in aquifer systems. It utilizes the finite-difference method to provide precise simulations (Ashraf & Ahmad, 2012).

It is also possible to replicate a wide range of characteristics and functions, including evapotranspiration, recharge from irrigation and precipitation, rivers, streams, drains, springs, reservoirs, and wells (Figure 8.6(a)). Using this technique, a grid divides an aquifer system into rectangular blocks (Figure 8.6(b)). Each block in the grid is referred to as a "cell," and it is arranged in rows, columns, and layers. The user needs to enter the aquifer attributes for each cell in the aquifer system's volume. Furthermore, for cells that correspond to the features' locations, the user enters details about rivers, wells, and other outflow and inflow features.

It is challenging to manage the numerous, dissimilar, and massive datasets needed for groundwater models. Coordinating data gathering, offering thorough database operations, enabling methodical model parameter assignments, carrying out geographical analysis (such as spatial statistics) activities, and presenting model results in easily understood color map forms are just a few ways that GIS can assist with the modeling process (Dawoud et al., 2005).

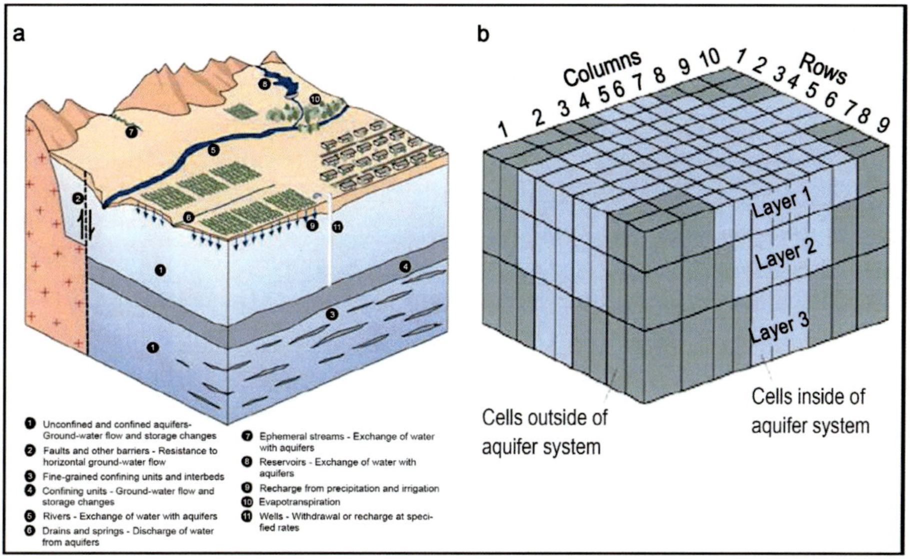

Figure 8.6. *Description of the (a) aquifer characteristics that MODFLOW can stimulate; and (b) the three-dimensional finite-difference grid represents the features.*

Source: M. Ayers; Creative Commons License.

Gridded data is commonly used to represent groundwater systems, allowing for efficient creation and visualization of spatial distributions during model pre- and post-processing. Grid functions simplify comparing and adjusting input data. Using a GIS makes it easier to visualize important attribute values in huge data sets since it allows the values to be shown in map formats (Jonkman, 2005).

Data is typically collected at specific points and then transformed into elevation surfaces using geo-statistical techniques. These surfaces represent the land surface and piezo-metric elevations. The creation of input datasets for MODFLOW involves the utilization of GIS grids, coverage, and shape files. These datasets encompass various aspects such as stratigraphy and hydrogeology, hydrogeologic parameters, initial conditions, and boundary conditions. Geographical data is utilized to depict rivers, drains, and wells. Geoprocessing is utilized to generate polygons that encompass distinct soil types, precipitation patterns, evapotranspiration rates, and land use characteristics. These polygons are subsequently combined to produce arrays representing recharge and evapotranspiration (Zhou & Li, 2011).

Groundwater data and modeling support have been moving toward a geo database approach in tandem with the surface water domain. Zeiler described the object-oriented ESRI geodatabase data model as being included with the ArcGIS® program.

The model consists of a feature dataset, feature classes, object classes, and geodatabase. An array of monitoring stations could be included in a feature dataset. The dataset consists of a feature class representing the locations of observation wells and an object class containing the water level data collected from all the stations. The model is maintained within a geodatabase that encompasses the feature datasets. A common coordinate system is used to link the many feature classes found in the feature databases (Babiker et al., 2007).

Object classes refer to non-spatial entities such as data tables, while feature classes encompass objects along with their corresponding spatial coordinates. To include groundwater Maidment et al. extended the principles of the ArcHydro geo database.

Groundwater applications cover a broad spectrum of research, ranging from regional studies on aquifer flow to detailed investigations of the three-dimensional flow within specific sites. Figure 8.7 presents a clear visual depiction of the different categories and connections in the three-dimensional data model.

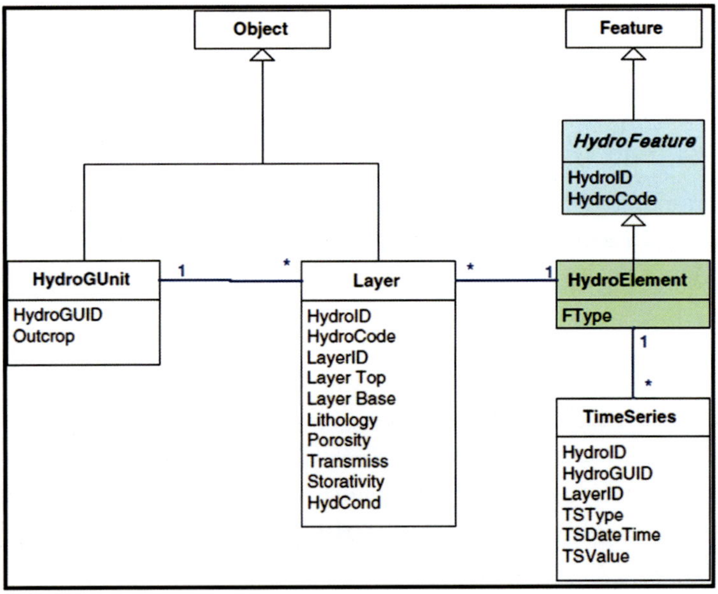

Figure 8.7. *Illustration of relationship and classes of the three-dimensional data model.*

Source: Stewart, R.; Creative Commons License.

8.3.3. GIS Use in Floodplain Management

8.3.3.1. Floodplain Mapping Requirements

Expanding the scope of surface water hydrologic modeling is crucial for accurate floodplain mapping, and the utilization of GIS tools is instrumental in achieving these objectives. Given the Federal Emergency Management Agency's (FEMA) regulatory jurisdiction, its guidelines are essential in establishing the fundamental mapping criteria for floodplains. FEMA offers comprehensive guidelines and specifications for the development of Flood Hazard Maps and other products related to the National Flood Insurance Program (NFIP). They also supervise the coordination and documentation activities associated with these efforts (Tate et al., 2002).

Did You Know?

GIS-based water supply demand forecasting enables utilities to anticipate future water needs based on factors such as population growth, industrial development, and climate change projections.

Three general categories of data are needed for floodplain mapping and management: Three types of data are needed for hydrologic and hydraulic modeling: (i) topography for the floodplain and watershed; (ii) physical data on buildings and constructed structures for drainage management; and (iii) administrative data on jurisdiction borders. These data are created from many sources via a range of GIS techniques and technologies, and they are eventually gathered into an extensive dataset that supports both short-term project requirements and long-term multiple-purpose management goals.

Using GIS techniques to update floodplain maps has become commonplace. The majority of current regulatory programs are based on floodplain maps and data that were created using conventional procedures in the 1970s and 1980s. Modernization programs are, and GIS guidelines are being released for them (Wolman, 1971).

Research conducted by the USGS on the Nisqually River near Puget Sound, Washington, showcased the success of utilizing contemporary elevation data and GIS technology to enhance flood maps. The current floodplain maps have several limitations – (i) using antiquated flood probability estimations; (ii) using difficult-to-manage hand-drawn procedures; (iii) having poor vertical accuracy; and (iv) the expensive and time-consuming nature of updates (Karr & Dudley, 1981). The study showed that the GIS approach had several advantages. Firstly, it was cost-effective, being only 10–20% of the cost of traditional methods. Secondly, it was just as accurate as traditional methods but provided more detailed information. Thirdly, it was able to provide data on the depth of floods. Fourthly, it could identify areas of uncertainty. Lastly, it was digital, which meant that it could be used to analyze other themes, such as buildings and roads, to support risk assessments. GIS was used to construct and modify digital elevation models of the flood surface and the land surface (Bradbrook et al., 2005).

Finding the flooded region is simple: subtract the land surface elevation model from the flood surface elevation model at each location. This produces negative results when the flood level is higher than the land elevation. This computation yields flood depth as a byproduct, which is significant for damage and insurance evaluations when it intersects with building floor elevations (Figure 8.8).

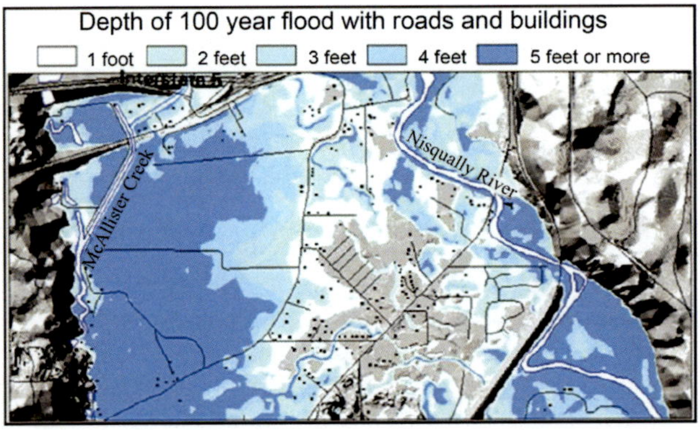

Figure 8.8. *Illustration of flood depth map buildings and roads of locations.*

Source: J. Kelman; Creative Commons License.

8.3.3.2. Floodplain Geo Database

Floodplain mapping requires a large variety of data sources, thus there's a strong need to aggregate the data into an extensive geodatabase. By doing this, one can benefit from the advantages that a geo database offers, such as standardization, elimination of redundancy, concurrency control, and transferability, among others. FEMA has created multiple data collecting and reporting criteria for floodplain studies as a component of its comprehensive effort to update maps across the country (Magliulo & Valente, 2020).

Regarding the creation of a geodatabase for FEMA flood hazard mapping, there are two key publications. Flood Hazard Mapping Partners' Guidelines and

Specifications, Appendix L: The first document is DFIRM Guidance and Draft Digital Data Database Preparation (Deshpande, 2013).

Appendix N of Flood Hazard Mapping Partners: Guidelines and Specifications: Data Capture Standards (DCS) is the second document. These guidelines describe the GIS databases that are currently being used to store flood hazard models and the associated data. The DCS relationship diagram for hydraulics is shown in Figure 8.9.

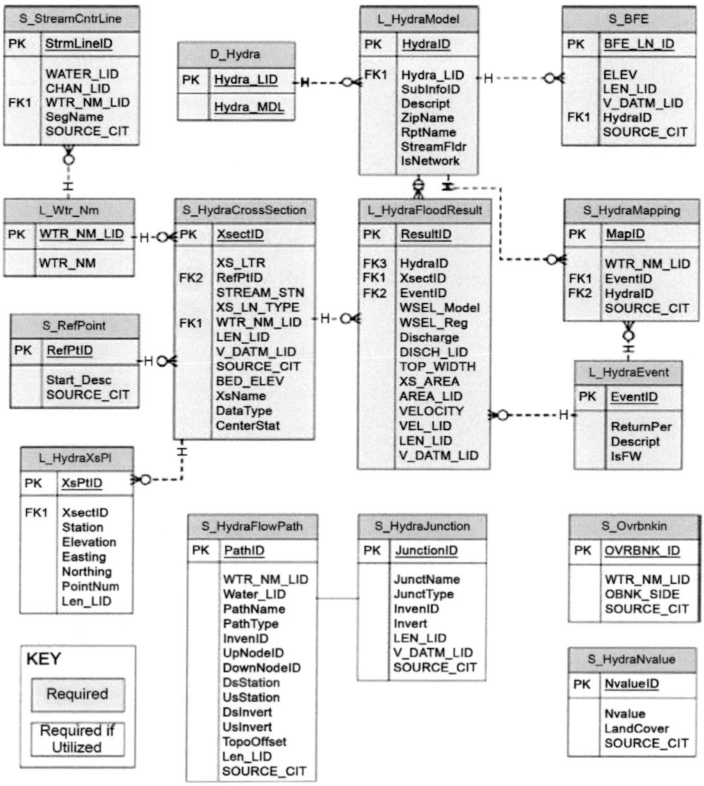

Figure 8.9 Illustration of Data Capture Standards relationship diagram for hydraulics (Source: M Ayers, Creative Commons License)

8.3.3.3. Floodplain Hydraulic Modeling with GIS

The technical foundation of floodplain studies involves the use of hydrologic and hydraulic modeling techniques to determine the boundaries of the floodplain. GIS plays a crucial role in conducting modeling studies, allowing for the integration of different data sources, coordination of various models, and creation of detailed maps essential for supporting flood management strategies (Paiva et al., 2011).

Numerous floodplain modeling packages for hydrologic and hydraulic (H&H) have been created by the public and business sectors. Floodplain hydraulics can be represented using H&H software, which offers a variety of capabilities and levels of GIS integration. Within the 1-D approach, the focus is on water flow primarily happening in a single spatial dimension, following the course along the main river channel. The model shows the structure of the issue by utilizing cross-sections that are perpendicular to the centerline of the channel. These sections encompass both the channel itself and the adjacent floodplain (Nunes et al., 1998).

Two spatial dimensions of depth-averaged velocities and water level can be computed using various computational grid methods. These include finite volume, finite element, and finite difference methods. The 2-D models are well-suited for scenarios where floodwaters have the potential to overflow from main channels and spread across the land, like on alluvial fans (Walling & Webb, 1981).

SUMMARY

- The chapter "Climate Change Impact and GIS Applications in Water Resources" provides a thorough examination of the complex connection between climate change and water management, highlighting the crucial role of Geographic Information Systems (GIS) in tackling related challenges.
- This chapter explores the phenomenon and explains its causes, including greenhouse gases, radiative forcing, and natural occurrences such as volcanic eruptions and ocean currents.
- The chapter investigates how climate change affects water resources, analyzing phenomena including runoff, floods, droughts, snowmelt, and glacier melt, declining water quality, depletion of groundwater, and disturbances to ecosystems.
- The chapter shifts to examine how GIS is used in flood control and water resource management. It describes the function of GIS in water supply systems, including modeling and analysis of groundwater hydrology, demand predictions, and pipe network design.
- Additionally, it explores the practical applications of GIS in managing floodplains, highlighting the necessary steps for accurately mapping floodplains, creating geodatabases, and conducting hydraulic modeling to improve resilience against floods.

REVIEW QUESTIONS

1. Define climate change and explain how it differs from natural climate variability.
2. Describe the primary causes of climate change, both natural and anthropogenic. Discuss the role of greenhouse gases in climate change.
3. How does climate change affect runoff patterns in different regions? Provide examples.
4. Discuss the ways in which climate change contributes to the frequency and severity of floods.
5. Explain the impacts of climate change on drought conditions and their implications for water resources.
6. Describe the potential effects of climate change on groundwater resources.

7. Discuss how climate change can affect ecosystems that rely on specific water regimes.
8. Explain how GIS technology can be utilized for managing water supply systems. Provide specific examples of its applications.
9. Describe how GIS can be used in floodplain management to mitigate the impacts of floods. Include examples of GIS tools or models that are commonly used.

MULTIPLE CHOICE QUESTIONS

1. What is the primary driver of climate change?

a. Deforestation

b. Greenhouse gases

c. Ocean currents

d. Volcanic eruptions

2. Which of the following is NOT a greenhouse gas contributing to climate change?

a. Carbon dioxide

b. Methane

c. Nitrous oxide

d. Oxygen

3. Which natural process contributes to climate change by releasing greenhouse gases?

a. Photosynthesis

b. Volcanic eruptions

c. Ocean circulation

d. Soil erosion

4. How does climate change impact water resources?

a. Decreased precipitation

b. Increased flooding

c. Reduced snowmelt

d. All of the above

5. Which GIS application is commonly used for water supply systems?

a. GIS-based flood forecasting

b. GIS for groundwater modeling

c. GIS-based water supply demand forecasting

d. GIS for floodplain management

6. What is a key component of floodplain management using GIS?

a. Hydrological modeling

b. Soil analysis

c. Vegetation mapping

d. Traffic monitoring

Answers to Multiple Questions

1. (b); 2. (a); 3. (c); 4. (c); 5. (a)   6.(a)

REFERENCES

1. Ashraf, A., & Ahmad, Z. (2012). Integration of groundwater flow modeling and GIS. *Water Resources Management and Modeling, 131*, 132.
2. Aston, R. J. (1973). Tubificids and water quality: A review. *Environmental Pollution (1970), 5*(1), 1–10.
3. Ault, T. R. (2020). On the essentials of drought in a changing climate. *Science, 368*(6488), 256–260.
4. Ayad, A., Awad, H., & Yassin, A. (2018). Integrated approach for the optimal design of pipeline networks. *Alexandria Engineering Journal, 57*(1), 87–96.
5. Babiker, I. S., Mohamed, M. A., & Hiyama, T. (2007). Assessing groundwater quality using GIS. *Water Resources Management, 21*, 699–715.
6. Beck, M. B. (1987). Water quality modeling: A review of the analysis of uncertainty. *Water Resources Research, 23*(8), 1393–1442.
7. Berrang-Ford, L., Ford, J. D., & Paterson, J. (2011). Are we adapting to climate change? *Global Environmental Change, 21*(1), 25–33.
8. Bhateria, R., & Jain, D. (2016). Water quality assessment of lake water: A review. *Sustainable Water Resources Management, 2*, 161–173.
9. Braca, G., Bussettini, M., Ducci, D., Lastoria, B., & Mariani, S. (2019). Evaluation of national and regional groundwater resources under climate change scenarios using a GIS-based water budget procedure. *Rendiconti Lincei. Scienze Fisiche e Naturali, 30*, 109–123.
10. Bradbrook, K., Waller, S., & Morris, D. (2005). National floodplain mapping: Datasets and methods–160,000 km in 12 months. *Natural Hazards, 36*, 103–123.

11. Brundrett, M. (1991). Mycorrhizas in natural ecosystems. In A. H. Fitter, F. Martin, & R. F. Read (Eds.), *Advances in Ecological Research, 21*, 171–313. Academic Press.
12. Cahill, A. E., Aiello-Lammens, M. E., Fisher-Reid, M. C., Hua, X., Karanewsky, C. J., Yeong Ryu, H., & Wiens, J. J. (2013). How does climate change cause extinction? *Proceedings of the Royal Society B: Biological Sciences, 280*(1750), 20121890.
13. Costa, J. E., & O'Connor, J. E. (1995). gynomorphically effective floods. In L. Mayer & D. Nash (Eds.), *Geophysical Monograph-American Geophysical Union, 89*, 45–45.
14. Da Silva, J. M. C., & Wheeler, E. (2017). Ecosystems as infrastructure. *Perspectives in Ecology and Conservation, 15*(1), 32–35.
15. Dawoud, M. A., Darwish, M. M., & El-Kady, M. M. (2005). GIS-based groundwater management model for Western Nile Delta. *Water Resources Management, 19*, 585–604.
16. Deshpande, S. S. (2013). Improved floodplain delineation method using high-density LiDAR data. *Computer-Aided Civil and Infrastructure Engineering, 28*(1), 68–79.
17. Diaz, H. F. (1983). Drought in the United States. *Journal of Applied Meteorology and Climatology, 22*(1), 3–16.
18. Dietz, T., Shwom, R. L., & Whitley, C. T. (2020). Climate change and society. *Annual Review of Sociology, 46*, 135–158.
19. Edition, F. (2011). Guidelines for drinking-water quality. *WHO Chronicle, 38*(4), 104–108.
20. Elbeih, S. F. (2015). An overview of integrated remote sensing and GIS for groundwater mapping in Egypt. *Ain Shams Engineering Journal, 6*(1), 1–15.
21. Farooq, M., Wahid, A., Kobayashi, N. S. M. A., Fujita, D. B. S. M. A., & Basra, S. M. (2009). Plant drought stress: Effects, mechanisms, and management. In N. Ahmad (Ed.), *Sustainable Agriculture,* 153–188. Springer Netherlands.
22. Giordano, M. (2009). Global groundwater? Issues and solutions. *Annual Review of Environment and Resources, 34*, 153–178.
23. Gorde, S. P., & Jadhav, M. V. (2013). Assessment of water quality parameters: A review. *Journal of Engineering Research and Applications, 3*(6), 2029–2035.
24. Hannon, B. (1973). The structure of ecosystems. *Journal of Theoretical Biology, 41*(3), 535–546.

25. Hansen, J., Sato, M., & Ruedy, R. (1997). Radiative forcing and climate response. *Journal of Geophysical Research: Atmospheres, 102*(D6), 6831–6864.

26. Haywood, J., & Boucher, O. (2000). Estimates of the direct and indirect radiative forcing due to tropospheric aerosols: A review. *Reviews of Geophysics, 38*(4), 513–543.

27. Hertzberg, M., Siddons, A., & Schreuder, H. (2017). Role of greenhouse gases in climate change. *Energy & Environment, 28*(4), 530–539.

28. Inanloo, B., Tansel, B., Shams, K., Jin, X., & Gan, A. (2016). A decision aid GIS-based risk assessment and vulnerability analysis approach for transportation and pipeline networks. *Safety Science, 84*, 57–66.

29. Jacobides, M. G., Cennamo, C., & Gawer, A. (2018). Towards a theory of ecosystems. *Strategic Management Journal, 39*(8), 2255–2276.

30. Jakeman, A. J., & Hornberger, G. M. (1993). How much complexity is warranted in a rainfall-runoff model? *Water Resources Research, 29*(8), 2637–2649.

31. Jayarathna, L., Rajapaksa, D., Managi, S., Athukorala, W., Torgler, B., Garcia-Valiñas, M. A., & Wilson, C. (2017). A GIS-based spatial decision support system for analyzing residential water demand: A case study in Australia. *Sustainable Cities and Society, 32*, 67–77.

32. Jonkman, S. N. (2005). Global perspectives on loss of human life caused by floods. *Natural Hazards, 34*(2), 151–175.

33. Karr, J. R., & Dudley, D. R. (1981). Ecological perspective on water quality goals. *Environmental Management, 5*, 55–68.

34. Kiselychnyk, O., Bodson, M., & Werner, H. (2009). Overview of energy-efficient control solutions for water supply systems. *Transactions of Kremenchug State Polytechnic University, 3*(56), 40–46.

35. Konikow, L. F., & Kendy, E. (2005). Groundwater depletion: A global problem. *Hydrogeology Journal, 13*, 317–320.

36. Ledley, T. S., Sundquist, E. T., Schwartz, S. E., Hall, D. K., Fellows, J. D., & Killeen, T. L. (1999). Climate change and greenhouse gases. *Eos, Transactions American Geophysical Union, 80*(39), 453–458.

37. Magliulo, P., & Valente, A. (2020). GIS-Based geomorphological map of the Calore River floodplain near Benevento (Southern Italy) overflooded by the 15th October 2015 event. *Water, 12*(1), 148.

38. Masera, O., Ghilardi, A., Drigo, R., & Trossero, M. A. (2006). WISDOM: A GIS-based supply demand mapping tool for wood fuel management. *Biomass and Bioenergy, 30*(7), 618–637.

39. Merz, B., Blöschl, G., Vorogushyn, S., Dottori, F., Aerts, J. C., Bates, P., & Macdonald, E. (2021). Causes, impacts, and patterns of disastrous river floods. *Nature Reviews Earth & Environment, 2*(9), 592–609.
40. Merz, R., & Blöschl, G. (2003). A process typology of regional floods. *Water Resources Research, 39*(12), 132-145.
41. Mishra, A. K., & Singh, V. P. (2010). A review of drought concepts. *Journal of Hydrology, 391*(1–2), 202–216.
42. Montzka, S. A., Dlugokencky, E. J., & Butler, J. H. (2011). Non-CO_2 greenhouse gases and climate change. *Nature, 476*(7358), 43–50.
43. Mukhopadhyay, B., & Khan, A. (2015). A reevaluation of the snowmelt and glacial melt in river flows within the Upper Indus Basin and its significance in a changing climate. *Journal of Hydrology, 527*, 119–132.
44. Nunes Correia, F., Castro Rego, F., Da Grača Saraiva, M., & Ramos, I. (1998). Coupling GIS with hydrologic and hydraulic flood modeling. *Water Resources Management, 12*, 229–249.
45. Omani, N., Srinivasan, R., Karthikeyan, R., Reddy, V., & Smith, P. K. (2016). Impacts of climate change on the glacier melt runoff from five river basins. *Transactions of the ASABE, 59*(4), 829–848.
46. Omer, N. H. (2019). Water quality parameters. *Water Quality-Science, Assessments and Policy, 18*, 1–34.
47. Paiva, R. C., Collischonn, W., & Tucci, C. E. (2011). Large-scale hydrologic and hydrodynamic modeling using limited data and a GIS-based approach. *Journal of Hydrology, 406*(3–4), 170–181.
48. Panu, U. S., & Sharma, T. C. (2002). Challenges in drought research: Some perspectives and future directions. *Hydrological Sciences Journal, 47*(S1), S19–S30.
49. Paola, C., Twilley, R. R., Edmonds, D. A., Kim, W., Mohrig, D., Parker, G., & Voller, V. R. (2011). Natural processes in delta restoration: Application to the Mississippi Delta. *Annual Review of Marine Science, 3*, 67–91.
50. Patten, B. C. (1981). Environs: The superniches of ecosystems. *American Zoologist, 21*(4), 845–852.
51. Pielke Jr, R. A. (2004). What is climate change? *Energy & Environment, 15*(3), 515–520.
52. Pimm, S. L. (1984). The complexity and stability of ecosystems. *Nature, 307*(5949), 321–326.
53. Sapkota, M., Arora, M., Malano, H., Moglia, M., Sharma, A., George, B., & Pamminger, F. (2014). An overview of hybrid water supply systems in

the context of urban water management: Challenges and opportunities. *Water, 7*(1), 153–174.

54. Schaner, N., Voisin, N., Nijssen, B., & Lettenmaier, D. P. (2012). The contribution of glacier melts to streamflow. *Environmental Research Letters, 7*(3), 034029.
55. Singh, A., & Gosain, A. K. (2011). Climate-change impact assessment using GIS-based hydrological modeling. *Water International, 36*(3), 386–397.
56. Singh, P., & Kumar, N. (1997). Impact assessment of climate change on the hydrological response of a snow and glacier melt runoff dominated Himalayan River. *Journal of Hydrology, 193*(1–4), 316–350.
57. Sophocleous, M. (2002). Interactions between groundwater and surface water: The state of the science. *Hydrogeology Journal, 10*, 52–67.
58. Swanston, D. N. (1991). Natural processes. *American Fisheries Society Special Publication, 19*, 139–179.
59. Tate, E. C., Maidment, D. R., Olivera, F., & Anderson, D. J. (2002). Creating a terrain model for floodplain mapping. *Journal of Hydrologic Engineering, 7*(2), 100–108.
60. Tholin, A. L., & Keifer, C. J. (1960). Hydrology of urban runoff. *Transactions of the American Society of Civil Engineers, 125*(1), 1308–1355.
61. Tóth, J. (1999). Groundwater as a geologic agent: An overview of the causes, processes, and manifestations. *Hydrogeology Journal, 7*, 1–14.
62. Wada, Y., Van Beek, L. P., Van Kempen, C. M., Reckman, J. W., Vasak, S., & Bierkens, M. F. (2010). Global depletion of groundwater resources. *Geophysical Research Letters, 37*(20), 1–99.
63. Walling, D. E., & Webb, B. W. (1981). Water quality. In J. M. Bayley & C. A. May (Eds.), *British Rivers, Vol. 1*, (pp. 126–169). John Wiley & Sons.
64. Wang, X., & Xie, H. (2018). A review of applications of remote sensing and geographic information systems (GIS) in water resources and flood risk management. *Water, 10*(5), 608.
65. Wilhite, D. A., & Glantz, M. H. (1985). Understanding the drought phenomenon: The role of definitions. *Water International, 10*(3), 111–120.
66. Wolman, M. G. (1971). Evaluating alternative techniques of floodplain mapping. *Water Resources Research, 7*(6), 1383–1392.
67. Zhou, Y., & Li, W. (2011). A review of regional groundwater flow modeling. *Geoscience Frontiers, 2*(2), 205–214.

INDEX